The Theory of Scintillation
with Applications
in Remote Sensing

The Theory of Scintillation with Applications in Remote Sensing

Charles L. Rino

IEEE PRESS

A JOHN WILEY & SONS, INC., PUBLICATION

Published by John Wiley & Sons, Inc., Hoboken, New Jersey.

Published simultaneously in Canada.

MATLAB and Simulink are registered trademarks of The MathWorks, Inc. See www.mathworks.com/trademarks for a list of additional trademarks. **The MathWorks Publisher Logo identifies books that contain MATLAB® content. Used with permission. The MathWorks does not warrant the accuracy of the text or exercises in this book or in the software downloadable from** http://www.wiley.com/WileyCDA/WileyTitle/productCd-047064477X.html **and** http://www.mathworks.com/matlabcentral/fileexchange/?term=authorid%3A80973. **The book's or downloadable software's use or discussion of MATLAB® software or related products does not constitute endorsement or sponsorship by The MathWorks of a particular use of the MATLAB® software or related products.**

For MATLAB® and Simulink® product information, or information on other related products, please contact:

The MathWorks, Inc.
3 Apple Hill Drive
Natick, MA 01760-2098 USA
Tel: 508-647-7000
Fax: 508-647-7001
E-mail: info@mathworks.com
Web: www.mathworks.com

For general information on our other products and services or for technical support, please contact our Customer Care Department within the United States at (800) 762-2974, outside the United States at (317) 572-3993 or fax (317) 572-4002.

Wiley also publishes its books in a variety of electronic formats. Some content that appears in print may not be available in electronic formats. For more information about Wiley products, visit our web site at www.wiley.com.

Library of Congress Cataloging-in-Publication Data:

Rino, Charles.
   The theory of scintillation with applications in remote sensing / Charles Rino.
      p. cm.
   Summary: "This book not only presents a thorough theoretical explanation of scintillation, but it also offers a complete library of MATLAB codes that will reproduce the book examples" — Provided by publisher.
   ISBN 978-0-470-64477-5 (hardback)
   1. Scintillation spectrometry. 2. Remote sensing. I. Title.
   QC373.S7R56 2011
   621.36'78—dc22   2010040949

10 9 8 7 6 5 4 3 2 1

Dedicated to Dr. Walter A. Flood for his unwavering support of university and industry scientists collaboratively engaged in propagation and scattering research from 1981 to 1995.

CONTENTS

PREFACE

This book evolved from a seminar on scintillation I presented at the International School on Atmospheric-Ionospheric Remote Sounding and Modelling (ISAR) held at the National Central University of Taiwan in October 2008. It was particularly rewarding to speak at the opening session, which was a tribute to Chao-Han Liu who was retiring as president of the National Central University after decades of contributions and leadership in the fields of remote sensing and scintillation.

In preparing material for the seminar it became clear that a theory of scintillation distinct from well-established theories of propagation and scattering had been largely taken for granted. Theoretical analyses developed to explain random fluctuations in galactic radio emissions, trans-ionospheric radio signals, optical images, and ocean acoustics signals have been collected and reviewed. Moreover, scintillation as a nuisance in radio communications, optical imaging, radar, and sonar has stimulated numerous application-specific theoretical developments. These developments have been strongly influenced by the diverse observational and analytical methodologies specific to each field. However, a unified theory of scintillation that builds on the common underpinnings of this large body of published material had been neglected.

Dialogue that took place during the seminar series suggested that the development of a theory of scintillation was a worthwhile endeavor. The basic material is well established, for example in the review paper written in 1982 by Chao-Han Liu and his long-time colleague Kung Chie Yeh [1]. Experiments and observations that stimulated the developments are also well documented in the Yeh-Liu review and in the 1982 review by Jules Aarons [2], who initiated many of the ionospheric global scintillation observation programs that followed the launch of Sputnik in 1957.

The fact that the ISAR theme was "remote sounding," synonomous with remote sensing, is noteworthy. The weak-scatter theory used extensively in radar remote sensing applications accommodates both forward scatter and backscatter in a simple and intuitive way. Scintillation theory neglects backscatter altogether, but it can accommodate modification of the forward-propagating excitation wave field that is neglected in the weak-scatter theory. When the media structure itself generates significant backscatter, both the weak-scatter theory of remote sensing and the theory of scintillation break down. Under such extreme conditions the theory of radiative transport must be used. The applications of radiative transport theory to remote sensing have been surveyed by Akira Ishimaru in a review paper [3] and in a two-volume series on the entire subject of remote sensing [4] [5]. Thus, scintillation theory is uniquely situated between scatter theory and radiative transport.

The material presented in this book was influenced by more than a decade of satellite scintillation research that began with the 1976 launch of the P76-5 satellite [6]. I had the good fortune to start my professional career working on this program and related projects involving radar sounding of the ionosphere. I am deeply indebted to my teachers, my co-workers, and numerous colleagues who freely shared their ideas, their data, and their critiques of my work.

I am pleased to acknowledge the meticulous checking of the analysis presented in Chapters 3 and 4 by Charles Carrano at the Institute for Scientific Research, Boston College. The data used in Chapter 5 were provided with generous help by Robert Livingston at Scion Associates. Susan Hoover, a long-time colleague at Vista Research, Inc., provided invaluable help in proofreading and editing the manuscript.

All the book examples are reproducible with MATLAB codes that can be downloaded from the MATLAB Central File Exchange. The codes are organized into 9 independent MATLAB Central File Exchange downloads. Internet links for the examples grouped by book chapter can be found on the following websites:

http://www.wiley.com/WileyCDA/WileyTitle/productCd-047064477X.html
http://www.mathworks.com/matlabcentral/fileexchange/
?term=authorid%3A80973

CHAPTER 1

INTRODUCTION

Twinkle, twinkle, little star. How I wonder what you are.
—Jayne Taylor

Scintillation observations have been used to identify and diagnose irregular structure in highly varied propagation media. Scintillation research has led to contributions in atmospheric physics, ionospheric physics, geophysics, ocean acoustics, and astronomy. The collection of papers published in the book *Wave Propagation in Random Media (Scintillation)* [7] provides a sampling of these diverse applications of scintillation phenomenology. *Scintillation* can be defined formally as a random modulation imparted to propagating wave fields by structure in the propagation medium. A familiar example is the twinkling of stars in the night sky, which is caused by turbulence in the earth's atmosphere. Structure in a propagation medium manifests itself as changes in the refractive index. From the ray theory of propagation it is known that local refractive-index enhancements can focus bundles of rays that pass through them. The wave field intensity variations identified as scintillation can be explained by this ray-focusing mechanism. Thus, scintillation phenomena are well understood and widely exploited, but nonetheless it is difficult to find a definitive treatment of the theory of scintillation. Introductory propagation theory addresses the space time evolution of wave fields launched in homogeneous media. Scattering theory addresses secondary (scattered) wave fields that are initiated wherever propagating wave fields encounter abrupt changes (steep gradients) in the material properties of the propagation medium. Scintillation theory lies somewhere in between the conventional theories of propagation and scattering.

This book has two objectives. The first objective is to develop a theory of scintillation that characterizes propagation media that support scintillation and the propagation phenomena that ensue. The second objective is to exploit numerical simulations as a research analysis tool. Modern and widely available computational resources support the application of high-fidelity and high-resolution simulations. The insights gained through simulations would be very difficult to obtain from analytic results and experimental observations alone.

The theoretical development presented in Chapters 2, 3, and 4 proceeds from an electromagnetic (EM) engineering framework with examples specific to propagation in the Earth's atmosphere and ionosphere. However, the formalism can be adapted to acoustic wave propagation in water or solid matter. This introductory chapter summarizes established results from EM theory that are used extensively in the development of scintillation theory and its applications. The interpretation of scintillation as a modulation imparted to a wave field that would otherwise be characterized completely by conventional propagation theory is emphasized as a link to real-world measurements and as an entree to modern digital signal processing techniques.

Chapter 2 presents a rigorous development of scintillation theory as a natural extension of propagation in strictly homogeneous media. Propagation media that admit structure with spatial variations that do not change over wavelength-scale distances are referred to as *weakly inhomogeneous media*. The implicit small-gradient constraint supports a modified wave equation that incorporates the structure as a multiplicative interaction with the total wave field. The wave equation thus modified is the starting point for all theoretical developments that address propagation in weakly inhomogeneous media. However, scintillation theory requires an additional assumption that sets it apart from other propagation phenomena. To make this requirement explicit, the modified wave equation is rewritten as an equivalent pair of coupled first-order differential equations that individually characterize propagation in opposite directions designated as forward and backward. Scintillation theory is based on the *forward approximation*, which neglects backward propagating waves induced by the interaction of the wave field with the weakly inhomogeneous structure. The resulting *forward propagation equation* (FPE) provides the mathematical connection between wave field observables and the structure that induces the modulation defined as scintillation. In mathematical terms, the theory of scintillation is a characterization of the domain and range of the

FPE. The more familiar parabolic wave equation follows from the parabolic approximation to the propagation operator in the FPE.

Chapter 2 concludes with simulation examples that illustrate the broad range of propagation phenomena supported by the FPE, including beam propagation and propagation in highly refracting backgrounds. Beam propagation is introduced to demonstrate the evolution of a propagating wave field as launched by real devices. The beam wave field is formally a carrier onto which scintillation structure is imparted. The final example demonstrates weak and strong scintillation in power-law environments via simulations. Plane-wave excitation is introduced as a way to avoid the need to model source and background details. In effect, plane-wave results characterize the propagation-induced modulation directly, although an adjustment must be made to accommodate wavefront curvature. The wavefront curvature correction is described in Chapter 4, where the corrections are used explicitly.

Chapter 3 develops the statistical theory of scintillation, which is a special case of the broader theory of scintillation supported by the FPE. The statistical theory provides a formalism for calculating statistical measures of the complex field in terms of statistical measures of the in situ structure. The complex field measures include probability distributions, spatial, temporal, and frequency coherence functions, spectral density functions, and structure functions. The main development in Chapter 3 is the derivation of a set of differential equations that individually characterize the complex-field moments of all orders in terms of summations of phase structure functions. The moment equations are themselves first-order differential equations with a structure similar to the underlying FPE equation. Key results that strike a balance between complexity and practical utility are presented, but an attempt has been made to reference major published analytic treatments that have contributed to a fundamental understanding of strong scatter. For completeness, a purely phenomenological theory that has been effective for modeling the probability distribution of the intensity fluctuations under strong-scatter conditions is also included.

Chapter 3 concludes with simulations that illustrate the characteristics of extremely strong scatter in power-law environments. Simulations are important because the statistical theory is strictly applicable only to a subclass of weakly inhomogeneous media that admit a high degree of statistical homogeneity. For example, the background structure in which scintillation develops invariably contains mean refractive index profile variations induced by the large-scale thermodynamic or electromagnetic variables that trigger locally unstable regions. Wave fields propagating in media undergoing structure evolution are influenced by a continuum of structure scales, not all of which can be incorporated into a statistically homogeneous ensemble. It is unusual in nature to find the ideal conditions under which conventional model calculations apply. Nonetheless, the strong-scatter theory provides critical guidelines to develop robust measures that characterize the severity of scintillation and the attendant loss of temporal, spatial, and frequency coherence.

Robust measures of these metrics provide quantitative links to the underlying physics that cause the structure.

Chapter 4 extends the scintillation theory to accommodate beacon satellite measurements. Atmospheric structure has the basic characteristics of fluid turbulence in the equilibrium range, which is characterized by the well-known properties of Kolmogorov turbulence. The three-dimensional structure can be characterized by an isotropic spatial wavenumber spectrum with a fixed radial wavenumber power-law index. A single scale-free measure of turbulent strength characterizes the structure. Commercially available scintillation measurement systems are calibrated to report the structure constant. By comparison, the ionospheric structure that dominates satellite scintillation is anisotropic and the power-law spectral characteristics depend on processes that drive the structure evolution. Additionally, propagation distances are sufficiently large that even small oblique propagation angles cause large displacements of the central ray in the measurement plane. To accommodate oblique propagation, the FPE must be translated to a continuously displaced coordinate system that keeps the measurement plane centered on the principal propagation direction. In Chapter 4, the geometric transformations necessary to adapt the theoretical results of Chapter 3 to oblique propagation in anisotropic media are developed.

Chapter 4 also describes and illustrates computational utilities for calculating satellite orbits and the local mean magnetic field vector that represents the Earth's magnetic field. Chapter 4 concludes with fully three-dimensional simulations of satellite scintillation structure that are effectively measurement plane scans along the apparent drift direction. The simulation results are examples of structure characteristics that would be very difficult to infer from theoretical results alone. The simulations also demonstrate robust path-integrated measures that can be mapped to turbulent strength or, where appropriate, to a turbulent structure constant.

Chapter 5 develops a framework for applying FPE scintillation theory to systems performance analysis. This development is important for all aspects of scintillation analysis. The theoretical development in Chapters 2, 3, and 4 characterize time-harmonic carrier or reference fields. Slow time variations caused by motion of the propagation path relative to the disturbed structure are included, but systems that exploit wave propagation necessarily use waveforms with finite bandwidth. All such systems use some variant of a sequence of waveforms that upon reception undergo signal processing to extract information either encoded in the transmitted waveform sequence itself or imparted to the transmitted waveforms as information about the propagation medium or scattering objects therein. A channel model is introduced as the formal interface between propagation phenomenology and system performance. The channel model also incorporates noise as a fundamental performance limitation. The development establishes the critical dependence of system performance on the signal-to-noise ratio and the coherence over critical signal processing intervals.

Because of the crucial interplay between waveforms and signal processing procedures, the introduction of real scintillation data for measurement is withheld until Chapter 5. The vagaries of real data and robust measures are important considerations in bringing theory and data together for both system performance analysis and scientific research. No claims are made that all the issues have been resolved, but the development can help systems analysts and scientists achieve their objectives more efficiently. Chapter 5 concludes with a review of current beacon satellite applications.

Chapter 6 addresses the extension of the FPE scintillation theory to accommodate compact scattering objects and irregular boundary layers. Readers of this book are likely to be aware of the complexity and computational demands of these active areas of developmental research. Even so, practitioners of computational electromagnetics accept numerical simulations of the electromagnetic fields scattered from mathematical models of extremely complicated objects as definitive results. Limitations most often reside in either realization or measurement of electrical characteristics of the scattering surfaces, not the computational methods themselves. It is understood that extreme care must be taken to make sure the results are self-consistent and physically meaningful. The same is true of modeling propagation in the atmosphere and the ionosphere. Extending these results to fully accommodate a mix of scattering objects and irregular surfaces remains an open challenge.

Chapter 6 introduces scattering by compact objects separated by weakly inhomogeneous media. It is assumed that the scattering characteristics of the object in free space are known. An exact formulation for problems of this type builds on the theory of multiple scattering from collections of interacting scattering objects. Imposing the forward-scatter approximation greatly simplifies and effectively extends the results to objects embedded in weakly inhomogeneous scattering media. The two-way propagation problem that characterizes propagation effects of scintillation in backscatter radar applications is developed in detail, with examples that apply to synthetic aperture radar (SAR).

The most challenging extension of the theory, however, is accommodating boundaries, particularly irregular boundaries such as the Earth's oceans and terrain surfaces. Readers familiar with parabolic wave equation (PWE) methods will know that there is an extensive literature of PWE applications that include acoustic propagation in the ocean and electromagnetic propagation above terrain and ocean surfaces. The rigorous development of PWE methods is generally confined to cylindrical wave propagation and largely heuristic treatments of irregular surfaces. However, the PWE applications to surface scatter use the forward approximation. Furthermore, random phase changing screens are routinely incorporated in these applications to model atmospheric structure. The development in Chapter 6 uses boundary integral methods with simplification obtained by invoking the forward approximation. The conditions under which these simplifications can be used do not extend to the computation of bistatic scattering cross sections, but the development does provide a fully three-dimensional formulation of forward propagation at low grazing angles in weakly inhomogeneous media with an irregular boundary.

To increase the utility of this book as a reference, each chapter contains tables of the symbols and abbreviations that are introduced within the chapter. (See Tables 1.1 and 1.2.)

Table 1.1 Chapter 1 Symbols

Symbol	Definition
$\mathbf{E}$, $\mathbf{H}$	Electric, Magnetic field vectors (volts/m,amps/m)
$\mathbf{J}_m$, $\mathbf{J}_e$	Magnetic, Electric current vectors (volts/m^2, amps/m^2)
$\mathbf{r} = [x, \varsigma]$	Position vector [reference axis, transverse (vector) axis]
$\omega = 2\pi f$	Frequency (radians/sec) (Hertz)
ϵ	Electric permittivity (Farads/m mks)
μ	Magnetic permeability (Henries/m mks)
$\eta = \sqrt{\mu/\epsilon}$	Intrinsic impedance (Ohms)
$c = 1/\sqrt{\mu\epsilon}$	Propagation velocity (mps)
$k = 2\pi f/c$	Wavenumber magnitude
$\mathbf{k}^{\pm} = [\pm k_x(\kappa), \boldsymbol{\kappa}]$	Vector spatial wavenumber
$k_x(\kappa) = k\sqrt{1 - (\kappa/k)^2}$	Magnitude of x-component of $\mathbf{k}^{\pm}$
$\widehat{\mathbf{E}}(x; \kappa) = \iint \mathbf{E}(x, \rho)\exp\{-i\boldsymbol{\kappa}\cdot\varsigma\}d\varsigma$	Two-dimensional spatial Fourier decomposition
$\mathbf{u} = (-\kappa_y, \kappa_z)/\kappa$	Horizontal polarization vector
$\mathbf{v}^{\pm} = (-\kappa/k, \pm g(\kappa)\kappa/\kappa)$	Perpendicular polarization vector
$\theta = \arccos(k_x(\kappa)/k)$	Propagation angle from reference axis
$\phi = \arctan(\kappa_y/\kappa_z)$	Propagation azimuth angle in $\mathbf{yz}$ plane
$\psi(x, \varsigma) \Leftrightarrow \widehat{\psi}(x; \kappa)$	Complex scalar wavefield/spatial Fourier transform
$G(\mathbf{u_k})$	Antenna gain function (2.7)
$\gamma^{\pm\pm}_{e_s e_i}(\kappa; \kappa_i)$	Bistatic scattering function (1.31)
$\sigma^{\pm\pm}_{e_s e_i}(\kappa; \kappa_i)$	Bistatic radar cross section (1.33)
$\mathbf{F}(x, \zeta)$	Modulation vector (1.42)
SNR	Receiver input signal-to-noise power ratio (1.39)
CF	System SNR constant (1.41)
BW	System noise bandwidth

Table 1.2 Chapter 1 Abbreviations

Abbreviation	Definition
EM	Electromagnetic
SNR	Signal-to-noise ratio
RCS	Radar cross section
PWE	Parabolic wave equation

1.1 ELECTROMAGNETIC PROPAGATION THEORY

This section reviews the theory of EM wave propagation to set the stage for the development of scintillation theory. The review starts with the symmetric time-harmonic form of Maxwell's equations:

$$\nabla \times \mathbf{E} = i\omega\mu\mathbf{H} - \mathbf{J}_m \tag{1.1}$$

$$\nabla \times \mathbf{H} = -i\omega\epsilon\mathbf{E} + \mathbf{J}_e \tag{1.2}$$

The complex vectors $\mathbf{E}$ and $\mathbf{H}$ represent, respectively, electric (volts/meter) and magnetic (amperes/meter) fields. The complex vectors $\mathbf{J}_e$ and $\mathbf{J}_m$ represent, respectively, electric and magnetic source functions. Each vector quantity has an implicit spatial dependence and an explicit but unwritten time variation $\exp\{-i\omega t\}$, where $\omega = 2\pi f$ is the frequency in radians per second, and f is the frequency in Hertz. In this development the constitutive parameters μ and ϵ are complex scalars, and only the electric permittivity ϵ will deviate from its free-space value. Thus, the characteristics of ϵ completely define the propagation environment. Through Maxwell's equations the complete response of the fields initiated by local source functions can be characterized. Complete developments of Maxwell's equations and the material summarized below can be found in the textbooks [8], [9], [10], and [11].

In a source-free region $\mathbf{J}_e = \mathbf{J}_m = 0$. It then follows from (1.1) and (1.2) that

$$\nabla \times \nabla \times \mathbf{E} - \omega^2\mu\epsilon\mathbf{E} = 0. \tag{1.3}$$

Using the vector identity

$$\nabla \times \nabla \times \mathbf{E} = -\nabla^2\mathbf{E} + \nabla\nabla \cdot \mathbf{E} \tag{1.4}$$

to eliminate the curl operation, the following equivalent form of the vector wave equation is obtained:

$$\nabla^2\mathbf{E} + \omega^2\mu\epsilon\mathbf{E} = \nabla\nabla \cdot \mathbf{E}$$
$$= \nabla(\mathbf{E}\cdot\nabla \log \epsilon). \tag{1.5}$$

The second equivalence in (1.5) follows by applying the divergence operator to (1.2). The vector wave equation forms the basis of all the subsequent analysis in this book. Attacking the vector wave equation directly would yield a theory of wave propagation in unconstrained inhomogeneous media [12]. The theory of scintillation starts with the assumption that the gradients in the material properties of the propagation media of interest are small enough that the gradient term in (1.5) can be neglected. In a strictly homogeneous medium, ϵ is a constant. Critical results that come from the theory of propagation in strictly homogeneous media are now reviewed.

1.1.1 Freely Propagating Waves

Freely propagating waves can be defined formally as homogeneous solutions to the second-order differential equation

$$\nabla^2 \mathbf{E} + \omega^2 \mu\epsilon \mathbf{E} = 0. \tag{1.6}$$

Freely propagating waves are classified by the shape of their constant-phase surfaces. Spherical waves are intrinsically well-matched to the evolving fields generated by highly compact sources, ideally points in space. Plane waves approximate the local fields measured at large distances from the field sources where the local properties of the field are of interest.

To pursue this dichotomy formally, note that direct computation shows that the family of plane waves of the form

$$\mathbf{E}(\mathbf{r}) = \widehat{\mathbf{E}}(\mathbf{k}) \exp\{i\mathbf{k} \cdot \mathbf{r}\} \tag{1.7}$$

satisfy (1.6) as long as the wave number is confined to the constant value

$$k = |\mathbf{k}| = \omega\sqrt{\mu\epsilon} = \omega/c. \tag{1.8}$$

Hereafter, vectors will be indicated in bold while scalars will be in normal font. The parameter c is the wave propagation speed, which for EM waves is the speed of light. The family of wave vectors associated with the plane-wave solutions to (1.6) define a sphere of radius k as long as k is real. Because of the constant-amplitude constraint, the components of the wave vector are not independent. Taking x as a reference direction, a wave vector of constant amplitude k has the form

$$\mathbf{k}^{\pm} = (\pm k_x(\kappa), \kappa), \tag{1.9}$$

where the transverse component κ is purely real, and

$$\begin{aligned} k_x(\kappa) &= k\sqrt{1 - (\kappa/k)^2} \\ &= kg(\kappa). \end{aligned} \tag{1.10}$$

If k is purely real, then

$$g(\kappa) = \begin{cases} \sqrt{1 - (\kappa/k)^2} & \text{for } \kappa \le k \\ i\sqrt{(\kappa/k)^2 - 1} & \text{for } \kappa > k \end{cases} . \tag{1.11}$$

When $\kappa \le k$, the associated plane wave propagates without attenuation along the ray defined by the polar angles

$$\theta = \arccos(g(\kappa)) \tag{1.12}$$
$$\phi = \arctan(\kappa_y/\kappa_z). \tag{1.13}$$

When $\kappa > k$, $k_x(\kappa)$ is purely imaginary and the associated plane wave decays exponentially. These evanescent waves can be sustained only in the vicinity of physical boundaries.

1.1.1.1 Polarization   Maxwell's equations impose constraints on the plane-wave field vectors as well as on the associated propagation vectors. Substituting (1.7) into (1.1) and (1.2) shows that

$$\mathbf{k} \times \widehat{\mathbf{E}} = \omega\mu\widehat{\mathbf{H}}. \tag{1.14}$$

Furthermore $\widehat{\mathbf{E}} \cdot \widehat{\mathbf{H}} = 0$ and $\mathbf{k} \cdot \widehat{\mathbf{H}} = 0$, which implies that $\widehat{\mathbf{E}}$ and $\widehat{\mathbf{H}}$ are confined to a plane perpendicular to $\mathbf{k}$. This polarization plane is spanned by the orthonormal polarization vectors

$$\mathbf{u} = \frac{\mathbf{a}_x \times \mathbf{k}^\pm}{|\mathbf{a}_x \times \mathbf{k}^\pm|} = (0, -\kappa_y, \kappa_z)/\kappa, \tag{1.15}$$

and

$$\mathbf{v}^\pm = \frac{\mathbf{u} \times \mathbf{k}^\pm}{|\mathbf{u} \times \mathbf{k}^\pm|} = (-\kappa/k, \pm g(\kappa)\kappa/\kappa). \tag{1.16}$$

Thus, the vector $\widehat{\mathbf{E}}$ can be written as

$$\widehat{\mathbf{E}} = \varepsilon_\| \mathbf{u} + \varepsilon_\perp \mathbf{v}^\pm, \tag{1.17}$$

where $\varepsilon_\|$ and $\varepsilon_\perp$ are complex scalars representing the amplitude of the $\widehat{\mathbf{E}}$ components parallel to the yz plane and perpendicular to $\mathbf{k}^\pm$, respectively. The vector $\mathbf{v}^\pm$ is reflected when the propagation direction is reversed. By introducing the explicit time dependence of the plane wave component, $\widehat{\mathbf{E}} \exp\{-i\omega t\}$, one can show that the direction of $\widehat{\mathbf{E}}$ traces out an ellipse with linearly polarized projections onto the basis vectors. Orthogonal left- and right-circularly polarized basis vectors can be constructed as $(\mathbf{u} \pm i\mathbf{v}^\pm)/\sqrt{2}$.

It follows that EM sources impose a polarization state on the fields they initiate. Elementary field probes are sensitive to linear combinations of the polarization vectors. If the receptor element responds to a polarization orthogonal to the source polarization, no energy is transferred. Thus, a measurement of the polarization state requires a pair of well-calibrated antennas. The reflection properties of dielectric surfaces are sensitive to polarization as are the scattering characteristics of material objects. Ionized media support polarization-dependent modes as well. However, the scintillation theory development will use scalar interactions that do not change the polarization state of the fields. The results can be generalized, but the added notational complexity can obscure the essential elements of the scintillation theory.

1.1.1.2 Plane-Wave Decomposition Now consider the two-dimensional Fourier decomposition of an electric field in a plane that is perpendicular to the x axis, which is defined as

$$\widehat{\mathbf{E}}(x_m; \boldsymbol{\kappa}) = \iint \mathbf{E}(x_m, \boldsymbol{\rho}) \exp\{-i\boldsymbol{\kappa}\cdot\boldsymbol{\varsigma}\} d\boldsymbol{\varsigma}. \qquad (1.18)$$

An extension of the wave field to any point in space beyond the plane can be constructed as

$$\begin{aligned} \mathbf{E}(\mathbf{r}) &= \iint \widehat{\mathbf{E}}(x_m; \boldsymbol{\kappa}) \exp\{ik_x(\boldsymbol{\kappa}) |x - x_m|\} \\ &\times \exp\{i\boldsymbol{\kappa}\cdot\boldsymbol{\varsigma}\} \frac{d\boldsymbol{\kappa}}{(2\pi)^2}. \end{aligned} \qquad (1.19)$$

Direct substitution will verify that (1.19) satisfies (1.6).[1] It follows that every solution to the homogeneous wave equation can be represented as a superposition of plane waves defined in a specified reference plane. In general, $\widehat{\mathbf{E}}(x_m; \boldsymbol{\kappa})$ consists of two additive components that individually support propagation in the opposite directions with respect to the reference plane x_m. The notation $\widehat{\mathbf{E}}^{\pm}(x_m; \boldsymbol{\kappa})$ will be used when it is necessary to keep track of the field components. The positive sign identifies propagation along the positive x direction; the negative sign identifies propagation in the opposite direction.

1.1.1.3 Far-Field Approximation The range of $\boldsymbol{\kappa}$ is formally the entire real plane, whereas propagating waves are confined to the disk circumscribed by $|\boldsymbol{\kappa}| = k$. Although sub-wavelength sampling is required to satisfy local boundary conditions, only propagating waves, most often with $\kappa \ll k$, carry energy to the far field. This can be exploited by evaluating (1.19) in the limit as $r \to \infty$. The far-field limit is

$$\lim_{r\to\infty} \mathbf{E}(\mathbf{r}) = -ik_x(\boldsymbol{\kappa}_r)\widehat{\mathbf{E}}(x_m; \boldsymbol{\kappa}_r) \frac{\exp\{ikr\}}{2\pi r}, \qquad (1.20)$$

---

[1] Fourier transformed fields are indicated by the carat. $\mathbf{E} \Leftrightarrow \widehat{\mathbf{E}}$, with $\Leftrightarrow$ indicating a Fourier transformation relation.

where κ_r is the magnitude of the transverse wave vector that propagates in the direction of **r**. Computational details are reproduced in Appendix A.1. The result shows that at large distances from any compact source the field can be represented by a spherical wave with a vector modulation proportional to the plane wave component that propagates toward the observation point. The corresponding result in a two-dimensional cylindrically symmetric system is

$$\lim_{r \to \infty} \psi(x, z) = \sqrt{i k_x(\kappa_r)} \widehat{\psi}(x_m; \kappa_r) \frac{\exp\{ikr\}}{\sqrt{2\pi kr}}. \tag{1.21}$$

Here ψ represents the principal component of **E** or **H** depending on polarization.

To explore the far-field relation further, recall that $\mathbf{E} \times \mathbf{H}^*$ represents a power flux.[2] It follows that

$$\lim_{r \to \infty} r^2 |\mathbf{E} \times \mathbf{H}^*| \, d\Omega \tag{1.22}$$

represents the power radiated into the unit solid angle $d\Omega = \sin\theta d\theta d\phi$ about the direction defined by the polar angles θ and ϕ. If k is purely real, substituting from (1.20) and (1.14) into (1.22) shows that

$$\lim_{r \to \infty} r^2 |\mathbf{E} \times \mathbf{H}^*| \, d\Omega = \frac{1}{\eta} \left| \widehat{\mathbf{E}}(x_m; \kappa_r) \right|^2 |k \cos \theta_r|^2 \frac{d\Omega}{(2\pi)^2}, \tag{1.23}$$

where $\eta = \sqrt{\mu/\epsilon}$. The clumsy form stems from the fact that rectangular and spherical coordinates have been mixed. Using the Jacobian $k^2 d\Omega/d\kappa_r = 1/|\cos\theta_r|$, it follows that

$$\lim_{r \to \infty} r^2 |\mathbf{E} \times \mathbf{H}^*| \, d\Omega = \frac{\cos\theta_r}{\eta} \left| \widehat{\mathbf{E}}(x_m; \kappa_r) \right|^2 \frac{d\kappa_r}{(2\pi)^2}. \tag{1.24}$$

Thus, the far-field power flux is precisely the power flux carried by the plane wave component propagating in the direction of **r**. Conservation of energy demands that the total power crossing a remote hemisphere be no more than the total power delivered by the source.

1.1.1.4 Antennas The sources of the propagating fields have been implied thus far. Antennas are physical devices that convert EM fields generated by power sources into freely propagating waves. Good antenna design ensures that the radiating field transmits as much of the power generated as possible in a prescribed direction. However this is achieved, the antenna system for

---

[2]Dimensionally, this is volts/meter times amps/meter equals watts per unit area. The asterisk denotes the complex conjugate.

most applications can be characterized by a gain function defined as follows:

$$G(\mathbf{u_k}) = \frac{\lim\limits_{r \to \infty} r^2 \left| \mathbf{E} \times \mathbf{H}^* \right|}{P_T/(4\pi)}$$

$$= \frac{\cos\theta_r \left| \widehat{\mathbf{E}}(x_m; \boldsymbol{\kappa}_r) \right|^2}{P_T/(4\pi)}. \tag{1.25}$$

Note that the form of G shows that the antenna gain is proportional to the Fourier transform of the aperture field distribution. In practice the normalization is derived from the field itself, which is formally a calibration procedure.

The standard application of the antenna gain function follows from (1.20). The power flux P delivered at range r is

$$P = P_T \frac{G(\mathbf{a}_r)}{4\pi r^2}. \tag{1.26}$$

The polarization-dependent energy delivered to a remote point in the far field of the antenna could be measured by a small loop or dipole placed in the field perpendicular to the ray path to the source. To detect in the far field usually requires an antenna with some gain and a low-noise amplifier (LNA). A reciprocity argument shows that the gain pattern of an antenna used for reception is proportional to its capture area A_r. The proportionality constant is ideally $\lambda^2/(4\pi)$. Thus, an antenna with gain G has the capture area $A_r = \lambda^2 G/(4\pi)$. It is sometimes convenient to deal with the aperture field directly as an effective amplitude gain function. A phase variation can be imposed that amounts to implicit electronic steering. The relationship to array theory is straightforward, but will not be pursued further here.

1.1.2 Bistatic Scattering Functions

The far-field form of the radiation from a compact source lends itself naturally to a self-contained definition of the antenna gain function. Upon defining Fourier transforms of scattered and transmitted fields in planes that lie in an unobstructed homogeneous medium, the far-field relations can be used again to extend a field to any remote point beyond the reference planes. Consider first a compact object bounded by two planes just beyond its x coordinate extrema. Let the object be illuminated by fields that are well-approximated by incident plane waves with polarization $\boldsymbol{\epsilon}_i = e_u^i \mathbf{u} + e_v^i \mathbf{v}^\pm$ and direction $\mathbf{k}_i^\pm = (\boldsymbol{\kappa}_i, \pm k_x(\boldsymbol{\kappa}_i))$ that point from the source to the scatterer. Let

$$\boldsymbol{\epsilon}_s \cdot \widehat{\mathbf{E}}_s^{\pm\pm}(\boldsymbol{\kappa}; \boldsymbol{\epsilon}_i, \boldsymbol{\kappa}_i) \tag{1.27}$$

represent the scattered fields at each plane for the particular plane-wave excitation. The first superscript refers to the direction of the scattered wave. The second superscript refers to the incident wave. As with the incident polarization vector, $\boldsymbol{\epsilon}_s = e_u \mathbf{u} + e_v \mathbf{v}^\pm$ is a unit vector that represents the polarization state of the scattered wave. The differential power flux for a specified incident and scattered polarization state is

$$
\begin{aligned}
\delta P_\epsilon^{\pm\pm}(\kappa; \kappa_i, \epsilon_i) &= \lim_{r\to\infty} r^2 \left| \epsilon_s \cdot \mathbf{E}_s(\mathbf{r}^\pm) \times \mathbf{H}_s(\mathbf{r}^\pm) \right| \\
&= \left| k\cos\theta \epsilon_s \cdot \widehat{\mathbf{E}}_s^{\pm\pm}(\kappa; \kappa_i, \epsilon_i) \right|^2 .
\end{aligned}
\tag{1.28}
$$

Boundary integral methods (see Section 6.2) could be used to calculate the field scattered by a compact object illuminated by a plane wave with wave vector $\mathbf{k}^\pm$. The form of the scattering function here is obtained by performing two-dimensional Fourier transformations of the fields propagating respectively in the forward hemisphere and the opposite hemisphere. This defines $\widehat{\mathbf{E}}_s^{\pm\pm}(\kappa; \kappa_i, \epsilon_i)$, although it is usually presented as a function of the four polar angles that define κ_i and κ. The wave vector formulation provides a more direct connection to the rigorous meaning of the bistatic scattering function.

Pursuing this further, note that

$$
\begin{aligned}
\overline{P}_{\epsilon\epsilon_i}^\pm(\kappa_i) &= \iint_A \cos\theta \left[\left| \epsilon_s \cdot \widehat{\mathbf{E}}_s^{+\pm}(\kappa; \kappa_i, \epsilon_i) \right|^2 \right. \\
&\quad \left. + \left| \epsilon_s \cdot \widehat{\mathbf{E}}_s^{-\pm}(\kappa; \kappa_i, \epsilon_i) \right|^2 \right] \frac{d\kappa}{4\pi^2}
\end{aligned}
\tag{1.29}
$$

represents the total power flux collected with polarization ϵ_s for the indicated plane-wave excitation. The total power is obtained by summing the contributions from two orthogonally polarized receive antennas. For plane-wave illumination $\widehat{\mathbf{E}}_i^+(\kappa_i)$, the total scattered power cannot exceed

$$
P_{e_i}(\kappa_i) = \cos\theta_i \left| \epsilon_i \cdot \widehat{\mathbf{E}}_i^+(\kappa_i) \right|^2 .
\tag{1.30}
$$

Thus, a bistatic scattering function that depends only on the direction of the incident illumination can be defined as

$$
\begin{aligned}
\gamma_{ee_i}^{\pm\pm}(\kappa; \kappa_i) &= 4\pi\delta P_\epsilon^{\pm\pm}(\kappa; \kappa_i, \epsilon_i)/P_{\epsilon e_i}(\kappa_i) \\
&= \frac{\left| k\cos\theta \epsilon_s \cdot \widehat{\mathbf{E}}_s^\pm(\kappa; \kappa_i, \epsilon_i) \right|^2}{\cos\theta_i \left| \epsilon_i \cdot \widehat{\mathbf{E}}_i^\pm(\kappa_i) \right|^2} .
\end{aligned}
\tag{1.31}
$$

This definition satisfies the energy conservation property

$$
\iint \left[\gamma_{e_s e_i}^+(\kappa; \kappa_i) + \gamma_{e_s e_i}^-(\kappa; \kappa_i) \right] d\Omega \le 1.
\tag{1.32}
$$

These results conform to the standard definition of the bistatic scattering function obtained by normalizing $\lim_{r\to\infty} r^2 |\mathbf{E}_s(\mathbf{r}) \times \mathbf{H}_s(\mathbf{r})|$ to the incident power flux along the direction of $\mathbf{k}^\pm$. Because illumination by any compact source is locally a plane wave in the far field, the definition is both general and practical. A well-calibrated antenna and source effectively define $\epsilon_i \cdot \widehat{\mathbf{E}}_i^{\pm}(\boldsymbol{\kappa}_i)$ at a fixed distance and direction from the illumination source.

One problem with this definition is that normalization to the power crossing the reference planes, rather than to the power propagating perpendicular to the incident direction, does not satisfy reciprocity. Thus, it is more common to use the bistatic radar cross section

$$\sigma_{e_s e_i}^{\pm\pm}(\boldsymbol{\kappa};\boldsymbol{\kappa}_i) = 4\pi \cos\theta_i \gamma_{e_s e_i}^{\pm\pm}(\boldsymbol{\kappa};\boldsymbol{\kappa}_i), \tag{1.33}$$

which satisfies the reciprocity relation

$$\sigma_{e_s e_i}^{\pm\pm}(\boldsymbol{\kappa};\boldsymbol{\kappa}_i) = \sigma_{e_i e_s}^{\pm\pm}(-\boldsymbol{\kappa}_i;-\boldsymbol{\kappa}). \tag{1.34}$$

Following the same arguments as before,

$$\sigma_{e_s e_i}^{\pm\pm}(\boldsymbol{\kappa}_r;\boldsymbol{\kappa}_i)/\left(4\pi r^2\right) \tag{1.35}$$

represents a cross-sectional area times the power flux per steradian at a distance r from the scatterer.

When the bistatic scattering functions are described in terms of polar angles with proper extensions over the entire scattering sphere, the signs take care of themselves. It is also possible to translate the plane-wave (four-angle) form of the scattering functions into source and measurement position form, effectively Green functions. Both forms can be used to formulate systems of algebraic equations that fully accommodate multiple scattering among discrete objects or discrete objects and surfaces [13], [14], [15], [16]. This construct will be used in Chapter 6 to accommodate compact scattering objects in the scintillation theory.

1.2 ANTICIPATING SCINTILLATION THEORY

This final section reviews a standard computation of signal strength and the free propagation of wave fields generated and captured by EM systems designed for communication or remote sensing applications. Scintillation is introduced as a modulation to the otherwise freely propagating wave field. It effect, it sets up a framework for the transition from propagation in homogeneous media to propagation in weakly inhomogeneous media.

1.2.1 Received Signal Power

The cost of an EM measurement, communication, or surveillance system is usually driven by the requirement that a minimal amount of signal power relative to the background noise must be present for effective signal capture. The background noise level is established by the first amplifier in a well-designed receiver system. System design and performance evaluation require computation of the strength of received power from a target in free space at the bistatic range $r = r_1 + r_2$, where r_1 is the range from the transmit antenna phase center to the target phase center and r_2 is the range from the target phase center to the receive antenna phase center. Phase centers are reference points from which phase can be measured. If the transmitter delivered P_T watts over the waveform duration, the following formula applies:

$$P_r = P_T \left[\frac{G_T(\mathbf{u}_1)}{4\pi r_1^2} \right] \left[\frac{\sigma(\mathbf{u}_1, \mathbf{u}_2)}{4\pi r_2^2} \right] \frac{\lambda^2 G_R(\mathbf{u}_2)}{4\pi} L_{\text{system}}. \tag{1.36}$$

The distribution of 4π factors and the definitions of the gain and scattering functions are dictated by engineering convention.

The grouping of the parameters here looks asymmetric, but it has been done with a purpose. Using the concepts just reviewed, the leftmost bracketed factor represents the power flux in watts per unit area normal to the beam axis direction, which is denoted by the unit vector $\mathbf{u}_1$, at the target location $\mathbf{u}_1 r_1$. The function $G_T(\mathbf{u}_1)$ is the transmit antenna gain in the direction of the target. The radar cross section $\sigma(\mathbf{u}_1, \mathbf{u}_2)$ is also constructed so that the middle factor in square brackets represents the power flux at $\mathbf{u}_2 r_2$. The final term is simply the capture area of the receive antenna.[3] The factor L_{system} represents losses in the receiver and antenna systems. A broader interpretation of the terms in square brackets would characterize the two-way propagation first from the transmit antenna to the target followed by the subsequent propagation of the scattered field from the target to the receive antenna. Chapter 2 develops the first stage of this concept by incorporating forward propagation in a disturbed medium. The scattering aspects will be developed in Chapter 6.

1.2.2 Noise Power

Noise power is defined by the relation

$$N_{\text{noise}} = k_B T_K (BW)_{\text{noise}}, \tag{1.37}$$

where k_B is Boltzmann's constant, T_K is the Kelvin noise temperature, and $(BW)_{\text{noise}}$ is the bandwidth over which the noise is measured. At the output of a well-designed receiver the noise power is

$$N_{\text{noise}} = k_B T_K (BW)_{\text{noise}} (NF), \tag{1.38}$$

where NF is the noise enhancement factor of the first amplifier. The bandwidth here is the receiver processor bandwidth, which is usually larger than the waveform bandwidth. The signal-to-noise ratio can be calculated as

---

[3]The $\lambda^2/4\pi$ factor that affects this transformation is not rigorously correct for all antennas, but the differences are usually absorbed in the calibration process.

$$SNR = P_T \left[\frac{G_T(\mathbf{u}_1)}{4\pi r_1^2}\right] \left[\frac{\sigma_{\text{radar}}(\mathbf{u}_1, \mathbf{u}_2)}{4\pi r_2^2}\right] \frac{\lambda^2 G_R(\mathbf{u}_2)}{4\pi}$$

$$\times \frac{L_{\text{system}}}{k_B T_K (BW)_{\text{noise}}(NF)}. \tag{1.39}$$

Because of the random character of noise, SNR represents the average signal strength that would be measured against the noise background. Using the receiver noise level as a reference eliminates the need to specify absolute power levels.

1.2.3 System Constant

It is convenient to group the factors that are constant under normal radar operation to obtain an alternate form

$$SNR = CF \frac{G_T(\mathbf{u}_1)}{G_{T\,\text{max}}} \frac{G_R(-\mathbf{u}_2)}{G_{R\,\text{max}}} \sigma_{\text{radar}}(\mathbf{u}_1, \mathbf{u}_2) / \left(r_1^2 r_2^2\right), \tag{1.40}$$

where CF, the system constant factor, is defined as

$$CF = P_T \frac{G_{T\,\text{max}} G_{R\,\text{max}} \lambda^2}{(4\pi)^3} \frac{L_{\text{system}}}{k_B T_K (BW)_{\text{noise}}(NF)}. \tag{1.41}$$

The constant factor could be estimated by measuring the return intensity of a target with known cross section. Precise measurement requires knowledge of the antenna gain patterns and the location of the calibration source. With appropriate monitoring to ensure that constant operating conditions are maintained, SNR measurements from a target whose position relative to the source and receive antennas is known can be used subsequently to estimate the target strength in RCS units. Because the system constant defines the maximum SNR at a specified range for a scatterer of known cross section, the system constant is a figure of merit for the receiver system.

Note that the SNR as defined here applies prior to any signal processing. Processing gain, which can increase the effective SNR considerably, will be discussed in Chapter 5. Clutter and interference rejection are also important, as are the receiver dynamic range and the physical environment in which the receiver must operate. The process starts, however, with the input SNR as defined by (1.40). For the special case of direct propagation from source to receiver, the radar equation applies with $\sigma_{\text{radar}}(\mathbf{u}_1, \mathbf{u}_2) / \left(4\pi r_2^2\right) = 1$. The polarization dependence has been ignored here. The polarization dependence is formally introduced with two complex vector dot products, one for transmission to target incidence, and one for target scattering to reception. For point-to-point transmission, a single complex vector dot product accommodates transmit-receive polarization differences.

1.2.4   Propagation Disturbances

The factors in (1.39) that accommodate propagation thus far apply only to strictly homogeneous media. The modifications that will be introduced in Chapter 2 to accommodate media structure can be characterized by a modulation factor defined as

$$\mathbf{F}(x,\zeta) = \frac{2\pi i r}{k_x(K_r)} \frac{\mathbf{E}(x,\zeta)\exp\{ikr\}}{\max\left|\widehat{\mathbf{E}}(x_m;\kappa_r)\right|}. \tag{1.42}$$

The modulation factor $\mathbf{F}(x,\zeta)$ is the ratio of the field at a given location in the medium to the field that would be measured in the far field in the absence of structure. The reason for normalizing to the maximum value of the aperture field is to avoid the ill-defined ratio of two small numbers for real measurements. Additionally, refraction imposes a different spatial variation that effectively defines the main propagation path through the medium. This refraction itself is important and can convey critical information about the background medium. In a communication system, the modulation imparted to the signal conveys information. The additional modulation imposed by the propagation medium is often a nuisance. This aspect of the problem will be developed in Chapter 5, where noise limitations are treated explicitly.

CHAPTER 2

THE FORWARD PROPAGATION EQUATION

If I have seen further than others, it is by standing upon the shoulders of giants.
—Isaac Newton

This chapter develops the forward propagation equation (FPE) as the defining equation of scintillation theory. The theory is built on two assumptions. The first assumption constrains gradients in the propagation medium. The second assumption constrains backscatter induced by structure in the medium. The class of *weakly inhomogeneous media* that satisfies the first assumption excludes discontinuous boundaries. The absence of definitive boundaries separates scintillation theory from classical scattering theory. However, it is the second assumption that establishes the unique characteristics of scintillation theory. All scattering interactions that involve backscatter are excluded, whereby energy propagating in the forward hemisphere is conserved. The field structure, which can be dramatically different from the initiating source field, is comprised entirely of interacting forward propagating waves. Aside from absorption in the background medium, no energy is lost.

The FPE development starts with a transformation of the modified wave equation as developed below into an equivalent pair of coupled first-order differential equations. The coupled equations are equivalent to the vector wave equation, but characterize waves propagating in opposite directions explicitly. The development uses two-dimensional spatial Fourier decompositions as described in Chapter 1. In the reference coordinate system, the direction normal to the decomposition plane is the propagation reference axis. The direction of propagation along the positive reference axis is designated *forward propagation*. Propagation in the opposite direction is designated *backward propagation*. In an extended weakly inhomogeneous medium the choice of the coordinate system is arbitrary, but the nature of the particular environment being analyzed usually dictates a prudent choice of reference coordinate system. Propagation oblique to a defined layered structure is a special case that will be treated in Chapter 4.

The early theory of radiowave scintillation and backscatter evolved from seminal papers by Booker, Radcliffe, and Shinn [17], Booker and Gordon [18], Briggs and Parkin [19], Mercier [20], Bowhill [21], and Budden [22], [23]. All these developments were constrained by the weak-scatter approximation, but they did not neglect backscatter. Although it is not emphasized in these early papers, the salient element of the weak-scatter theory is that it neglects the change in the excitation field as it evolves. The means of correcting this defect is surprisingly straightforward, but it separates scintillation theory from the backscatter theory that is used extensively in remote sensing to characterize atmospheric and ionospheric backscatter. The source of this backscatter is small-scale structure, which has a negligible effect on the forward propagating waves in a weakly inhomogeneous medium.

Section 2.1 develops the FPE equation and its relationship to more familiar theoretical results, such as the weak-scatter theory and the parabolic approximation, which constrains the range of forward scattering angles such that ray-optics methods are applicable. Section 2.2 introduces the numerical solution to the FPE and presents several examples that illustrate the range of problems formally accommodated by the FPE. (See Tables 2.1 and 2.2 for symbols and abbreviations.)

Table 2.1   Chapter 2 Symbols

Symbol	Definition
$S(\mathbf{r}) = k\delta n(\mathbf{r})$	Structure source function
$\widehat{S}(\kappa)$	3-D spatial wavenumber spectrum of $S(\mathbf{r})$ (2.9)
$\Delta\kappa^{\pm}(\kappa, \kappa')$	Bragg wavenumber (2.10)
$\widehat{S} \otimes \widehat{\mathbf{E}}(x; \kappa)$	Induced source wavenumber spectrum (2.12)
$G(k\,|\mathbf{r} - \mathbf{r}'|)$	Scalar Green function (2.7)
$\mathbf{E}^{\pm}(x, \varsigma) \Leftrightarrow \widehat{\mathbf{E}}^{\pm}(x; \kappa)$	Forward/Backward field components (2.14) (2.15)
$\ominus$	Spatial-domain propagation operator (2.18)
I	Identity operator
$\mathbf{s},\ \varkappa = d\mathbf{s}/ds$	Ray tangent, curvature (2.26)
$\tau_s,\ r_s$	Path delay, Path length (2.27), (2.28)
SI	Scintillation index (2.45)

Table 2.2   Chapter 2 Abbreviations

Abbreviation	Definition
FPE	Forward propagation equation
DFT	Discrete Fourier transformation
SDF	Spectral density function
SI	Scintillation index

2.1   WEAKLY INHOMOGENEOUS MEDIA

Section 1.1 developed the general form of the wave equation for inhomogeneous media, which is repeated here for reference:

$$\nabla^2 \mathbf{E} + k^2 n^2 \mathbf{E} = \nabla \left(\mathbf{E} \cdot \nabla \log \epsilon \right). \tag{2.1}$$

The factor $\omega^2 \mu \epsilon$ has been replaced by $k^2 n^2$ where $n = c/c_0$, with $c = 1/\sqrt{\mu\epsilon}$ the local wave velocity in the medium and c_0 the velocity of light in a vacuum. Thus $k = \omega/c_0$. Weakly inhomogeneous media admit structure with gradi-. ents constrained so that the right-hand side of (2.1) can be approximated by zero. Refractive index variations are introduced as perturbations to the local refractive index[4]

$$n = (1 + \delta n).$$

The defining equation for propagation in weakly inhomogeneous media now can be rewritten to isolate the structure term as a multiplicative interaction with the total field:

$$\begin{aligned} \nabla^2 \mathbf{E} + k^2 \mathbf{E} &= -k^2 (2\delta n + \delta n^2) \mathbf{E} \\ &\simeq -2k^2 \delta n \mathbf{E}. \end{aligned} \tag{2.2}$$

Reiterating the critical definitions, $k = 2\pi f/c_0$ with the understanding that k is an invariant wavenumber representing a vacuum background reference. The perturbation δn incorporates the spatial variation in its entirety. The constituent contributions to δn and their sources will be discussed in detail in Chapter 4. However, for the FPE development it is necessary only that δn satisfy the weakly homogeneous media small-gradient constraint.

2.1.1   Integral-Equation Form

The standard method of solving the wave equation in the presence of im-pressed or induced sources uses an equivalent integral form based on Green functions. The development of the FPE starts there as well, which provides continuity with alternative methods of solving the wave equation in weakly inhomogeneous media. The equivalent integral-equation form of (2.2) is con-structed by using the scalar free-space Green function

$$G(\mathbf{r}, \mathbf{r}') = \frac{\exp\left\{ik\left|\mathbf{r} - \mathbf{r}'\right|\right\}}{2\pi\left|\mathbf{r} - \mathbf{r}'\right|}, \tag{2.3}$$

which has the formal property

$$\nabla^2 G(\mathbf{r}, \mathbf{r}') + k^2 G(\mathbf{r}, \mathbf{r}') = -\delta\left(\mathbf{r} - \mathbf{r}'\right), \tag{2.4}$$

---

[4]In nonmagnetic media $n = \sqrt{\epsilon/\epsilon_0}$, whereby $n = \sqrt{1 + \delta\epsilon/\epsilon_0}$ and $\delta n \simeq \delta\epsilon/(2\epsilon_0)$.

where $\delta\left(\mathbf{r} - \mathbf{r}'\right)$ is the Kronecker delta function. By using the formal property of the delta function,[5] direct substitution will verify that the following scatter integral equation satisfies (2.2):

$$\mathbf{E}(\mathbf{r}) = \mathbf{E}_i(\mathbf{r}) + 2k \iiint G(\mathbf{r}, \mathbf{r}') S(\mathbf{r}') \mathbf{E}(\mathbf{r}') dV, \qquad (2.5)$$

where

$$S(\mathbf{r}) = k\delta n(\mathbf{r}). \qquad (2.6)$$

The vector field $\mathbf{E}_i(\mathbf{r})$ is a solution to the homogeneous wave equation. In principle, (2.5) could be solved directly for the unknown field $\mathbf{E}(\mathbf{r})$, but series solutions based on successive approximations to $\mathbf{E}(\mathbf{r})$ initiated by $\mathbf{E}_i(\mathbf{r})$ are tedious to implement and convergence is often poor.

2.1.2 Weak-Scatter Approximation

The objective here is to develop a linear integral relation between the incident source field and the scattered field. The approach follows the development in Rino and Fremouw [24]. The critical step is to replace $\mathbf{E}(\mathbf{r}')$ in the integrand of (2.5) by a *freely propagating* wave field that can be represented by (1.19). This is the essential element of the weak-scatter approximation, namely that the excitation field is unaffected by its interaction with the media structure.

The Wyle representation

$$G(k\left|\mathbf{r} - \mathbf{r}'\right|) = \iint \frac{i \exp\{ikg(\kappa)|x - x'|\}}{2kg(\kappa)} \exp\{i\boldsymbol{\kappa} \cdot \Delta\boldsymbol{\zeta}\} \, d\boldsymbol{\kappa}/(2\pi)^2, \qquad (2.7)$$

is substituted for the Green function in (2.5) [12, Section 2.2]. Implicit integration ranges are $-\infty$ to ∞. After some manipulation the following equivalent form is obtained:

$$\begin{aligned}
\mathbf{E}_s(\mathbf{r}) &= \mathbf{E}(\mathbf{r}) - \mathbf{E}_i(\mathbf{r}) \\
&= i \iint \left[\iint \widehat{S}(\Delta\kappa^\pm\left(\boldsymbol{\kappa}, \boldsymbol{\kappa}'\right)) \widehat{\mathbf{E}}(0; \boldsymbol{\kappa}') \frac{d\boldsymbol{\kappa}'}{(2\pi)^2} \right] \\
&\quad \times \frac{\exp\{\pm ikg(\kappa)x\}}{g(\kappa)} \exp\{i\boldsymbol{\kappa} \cdot \boldsymbol{\zeta}\} d\boldsymbol{\kappa}/(2\pi)^2.
\end{aligned} \qquad (2.8)$$

The function $\widehat{S}(\varkappa; \mathbf{K})$ is defined as

$$\widehat{S}(\varkappa; \mathbf{K}) = \int_0^{l_p} \iint S(x', \boldsymbol{\zeta}') \exp\{-i\varkappa x\} \exp\{-i\mathbf{K} \cdot \boldsymbol{\zeta}'\} \, dx' d\boldsymbol{\zeta}', \qquad (2.9)$$

[5] $\iiint \delta\left(\mathbf{r} - \mathbf{r}'\right) f(\mathbf{r}') d\mathbf{r}' = f(\mathbf{r}).$

which is the three-dimensional spatial wavenumber spectrum of the structure in the volume bounded by planes at $x = 0$ and $x = l_p$. The semicolon indicates the special treatment of the x variable. However, as $\widehat{S}(\varkappa; \mathbf{K})$ is evaluated in (2.8) it depends only on the transverse wavenumber variables $\boldsymbol{\kappa}$ and $\boldsymbol{\kappa}'$. The sign of $kg(\kappa')$ depends on where $\mathbf{r}$ lies relative to the boundary planes:

$$\Delta\boldsymbol{\kappa}^{\pm}(\boldsymbol{\kappa}, \boldsymbol{\kappa}') = \begin{cases} (kg(\kappa), \boldsymbol{\kappa}) - (kg(\kappa'), \boldsymbol{\kappa}') & \text{for } x > l_p \\ (kg(\kappa), \boldsymbol{\kappa}) - (-kg(\kappa'), \boldsymbol{\kappa}') & \text{for } x < 0 \end{cases} . \qquad (2.10)$$

Allowing for the location of the structure relative to the observation point, it follows that $\Delta\boldsymbol{\kappa}^{\pm}(\boldsymbol{\kappa}, \boldsymbol{\kappa}')$ is the difference between the incident and scattered wave vectors. It is usually referred to as the Bragg wavenumber, which plays a special role in remote sensing applications that exploit backscatter ($x < 0$) from smallscale structure in the propagation medium. A similar expression applies to the backscatter from small amplitude surface roughness.

The x variable defines the reference axis for propagation in the development of the theory. The propagation disturbance in the region $x > l_p$ develops under forward propagation conditions. Scattered fields in the region $x < 0$ come from backscatter. The term in square brackets in (2.8) can be interpreted as a spatial wavenumber spectrum obtained by convolving the incident wave spectrum with the spectrum defined by (2.9). The multiplicative terms in (2.8) that follow the square-bracketed term account for free-space propagation of the waves away from the disturbed layer. The difference between the scintillation and backscatter forms of the weak-scatter theory as presented here is the wavenumber range of the irregularity spectrum selected by the corresponding forms of (2.10). When $x > l_p$, the upper sign selects the largescale structure that supports scintillation. When $x < 0$, the lower sign selects the small scale structure that supports basckscatter.

In the usual implementation of weak-scatter theory the excitation field is replaced by a single incident plane wave. The incident wave spectrum is formally a delta function centered on the incident wave direction, which eliminates the convolution. The surviving modulation is the spectral content of the structure evaluated at the Bragg wavenumber defined by (2.10). In Appendix A.2 the form of (2.8) that is used for remote sensing applications that exploit backscatter from small-scale structure is developed. The applications include atmospheric backscatter from turbulent layers and ionospheric backscatter from structured layers. A variety of phenomena give rise to detectable backscatter. Equatorial plumes are among the most dramatic [25, Chapter 4].

Reiterating, the limitation of the weak-scatter theory of scintillation stems from the fact that the changes in the wave field due to interaction with the structured medium are ignored. Necessary conditions for the weak-scatter theory are difficult to establish, but sufficiency usually can be checked by direct computation. The importance of (2.8) for remote sensing is that it establishes a linear integral relation between an observable *scattered* field $\mathbf{E}_s(\mathbf{r})$

and the in situ structure that is initiating the scatter or the propagation disturbance. However, identifying scintillation-induced structure as a scattering phenomenon has been avoided in this book to emphasize the unique characteristics of scintillation theory.

2.1.3   Forward Approximation

The critical step in the development of an unrestricted theory of scintillation is to account for the modification of the evolving total field as the structure develops. The development is taken from Rino and Kruger [26]. Transformation of (2.5) to the spatial Fourier domain leads to the equivalent form

$$\widehat{\mathbf{E}}(x;\kappa) = \widehat{\mathbf{E}}_i(x;\kappa)$$
$$+ \int_{x_0}^{l_p} \frac{i\exp\{ikg(\kappa)|x-x'|\}}{g(\kappa)} \widehat{S} \otimes \widehat{\mathbf{E}}(x';\kappa)dx', \quad (2.11)$$

where $g(\kappa) = k\sqrt{1-(\kappa/k)^2}$ was introduced in Section 1.1. The symbol $\otimes$ is used to represent the integral product operation

$$\widehat{S} \otimes \widehat{\mathbf{E}}(x;\kappa) = \iint S(x,\varsigma')\mathbf{E}(x,\varsigma')\exp\{-i\kappa\cdot\varsigma'\}d\varsigma'. \quad (2.12)$$

Now let
$$\widehat{\mathbf{E}}(x;\kappa) = \widehat{\mathbf{E}}^+(x;\kappa) + \widehat{\mathbf{E}}^-(x;\kappa), \quad (2.13)$$

where $\widehat{\mathbf{E}}^+(x;\kappa)$ and $\widehat{\mathbf{E}}^-(x;\kappa)$ are defined as follows:

$$\widehat{\mathbf{E}}^+(x;\kappa) = \widehat{\mathbf{E}}_i(x;\kappa) \quad (2.14)$$
$$+ \frac{i\exp\{ikg(\kappa)x\}}{g(\kappa)} \int_{x_0}^{x} \exp\{-ikg(\kappa)x'\}\widehat{S} \otimes \widehat{\mathbf{E}}(x';\kappa)dx',$$

and

$$\widehat{\mathbf{E}}^-(x;\kappa) = \frac{i\exp\{-ikg(\kappa)x\}}{g(\kappa)} \int_{x}^{l_p} \exp\{ikg(\kappa)x'\}\widehat{S} \otimes \widehat{\mathbf{E}}(x';\kappa)dx'. \quad (2.15)$$

The forward component designated by the plus superscript admits scatter contributions only from source structure preceding the observation point x while the backward component, designated by the minus superscript, admits scatter contributions only from structure beyond the observation point. Upon computing the derivatives of (2.14) and (2.15) with respect to the variable x, the following coupled first-order differential equations are obtained:

$$\pm\frac{\partial\widehat{\mathbf{E}}^\pm(x;\kappa)}{\partial x} = ikg(\kappa)\widehat{\mathbf{E}}^\pm(x;\kappa) + \frac{i\widehat{S} \otimes \widehat{\mathbf{E}}(x;\kappa)}{g(\kappa)}. \quad (2.16)$$

Transforming these spectral-domain equations back to the spatial domain gives the most general spatial-domain form of the propagation equations for weakly inhomogeneous media, namely

$$\frac{\partial \mathbf{E}^{\pm}(x,\varsigma)}{\partial x} = \pm ik\Theta\mathbf{E}^{\pm}(x,\varsigma) + 2k \iint G(k\,|\varsigma - \varsigma'|)S(x,\varsigma')\mathbf{E}(x,\varsigma')d\varsigma'. \quad (2.17)$$

The propagation operator,

$$\Theta = \sqrt{1 + \nabla_{\perp}/k^2}, \quad (2.18)$$

is defined formally by the Taylor series expansion of $g(\kappa)$ and the equivalence between spatial-domain differentiation and frequency-domain multiplication by powers of spatial wavenumber components. However, at this point no approximation has been made in the derivation of (2.16).

The term involving $S(x,\varsigma')$ can be simplified first by noting that under the weak inhomogeneity constraint, $S(x,\varsigma')$ is a smooth function of ς' when compared to the rapid variation of the Green function term in the integrand. The field term that multiplies $S(x,\varsigma')$ generally would not acquire rapid variations with ς', whereby the source and field terms can be evaluated at ς and moved outside the integral in (2.17). The remaining integration of the Green function is readily identified as the zero-frequency term in the Fourier transformation. That is,

$$\iint G(k\,|\varsigma - \varsigma'|)d\varsigma' = i/(2k). \quad (2.19)$$

Thus, for almost all practical applications the simpler product form of (2.17) is used:

$$\frac{\partial \mathbf{E}^{\pm}(x,\varsigma)}{\partial x} = \pm ik\Theta\mathbf{E}^{\pm}(x,\varsigma) + iS(x,\varsigma)\mathbf{E}(x,\varsigma). \quad (2.20)$$

Transforming (2.20) to the spatial Fourier domain is equivalent to replacing $g(\kappa)$ in the media-interaction term of (2.16) by unity.

The structure of (2.20) suggests a forward-backward recursion, which has been used to accommodate backscatter in scintillation theory [27], but convergence beyond the first iteration is poor. The forward approximation is obtained by substituting the forward field for the total field in the media-interaction term. The following first-order differential equation defines the forward propagation equation

$$\frac{\partial \mathbf{E}^{+}(x,\varsigma)}{\partial x} = ik\Theta\mathbf{E}^{+}(x,\varsigma) + ik\delta n(x,\varsigma)\mathbf{E}^{+}(x,\varsigma). \quad (2.21)$$

The media-interaction term in (2.21) is a product in the spatial domain, whereas the propagation operator, $ik\Theta$, becomes a product in the Fourier domain. This suggests alternating between the spatial and the Fourier domains. In the absence of the propagation term, the solution takes the particularly simple form

$$\mathbf{E}^{+}(x_{n+1},\varsigma) = \mathbf{E}^{+}(x_n,\varsigma)\exp\left\{ik \int_{x_n}^{x_{n+1}} \delta n(x',\varsigma)dx'\right\}. \quad (2.22)$$

This motivates the split-step method in which the phase perturbation (2.22) is applied at the layer entrance followed by the propagation operator. Alternatively, (2.21) could be solved by finite-difference methods, which requires a means to approximate the propagation operator. The simplest approximation to the diffraction operator leads to the parabolic wave equation.

2.1.4   Parabolic Wave Equation

The parabolic approximation to the scalar wave equation is obtained by truncating the formal Taylor series expansion of the propagation operator at the leading term:

$$ ik\Theta \simeq ik + i\nabla_\perp / (2k). \tag{2.23} $$

This constrains the transverse field derivatives, which means the propagating field is confined to a narrow cone of propagation angles. More elaborate approximations extend the range of scattering angles that are accommodated in the numerical solution, but the parabolic approximation is used most extensively. Applying the parabolic approximation to (2.21) leads to the parabolic wave equation (PWE)

$$ 2k\frac{\partial\left[\mathbf{E}^+(x,\varsigma)\exp\{-ikx\}\right]}{\partial x} = i\nabla_\perp\left[\mathbf{E}^+(x,\varsigma)\exp\{-ikx\}\right] + $$
$$ 2ik^2\delta n(x,\varsigma)\left[\mathbf{E}^+(x,\varsigma)\exp\{-ikx\}\right]. \tag{2.24} $$

The term in square brackets is usually identified by a separate symbol. This is equivalent to replacing Θ by $\Theta - I$. In the PWE literature, (2.21) is called the wide-angle PWE, although it is more appropriate to call the PWE the narrow-angle approximation to

$$ \frac{\partial\psi(x,\varsigma)}{\partial x} = ik\left[\Theta - I\right]\psi(x,\varsigma) + ik\delta n(x,\varsigma)\psi(x,\varsigma). \tag{2.25} $$

Because no polarization change is induced by field interactions in weakly inhomogeneous scalar media, allowing the scalar $\psi(x,\varsigma)$ to represent any component of $\mathbf{E}^+(x,\varsigma)\exp\{-ikx\}$ imposes no loss of generality.

The advantage of the parabolic approximation and its variants is the elimination of the Fourier transformations, which require a large support space. Direct numerical solutions can be confined to a compact evolving region of interest. This is important in ocean acoustics, for example, where propagation may proceed over thousands of kilometers [28].

2.1.5   Ray Optics

Ray optics is an important complement to scintillation theory in that it provides a purely geometric construction that identifies the principal energy paths. Ray trajectories are defined formally by position-dependent vectors

$r(s)$, where s is the distance along the ray path. The vector $r(s)$ is constrained locally by two orthogonal vectors, namely the unit vector s tangent to the ray at $r(s)$ and the curvature vector $\varkappa = ds/ds$, which is normal to s. If changes in refractive index $n(r)$ over many wavelengths are small, the ray trajectory in the medium must satisfy the ray equation [29, Section 3.2][6]

$$n\varkappa + \frac{dn}{ds}s = \nabla n. \tag{2.26}$$

The ray equation is solved by initiating a ray with a prescribed starting position and direction and then using (2.26), with knowledge of the refractive index, to determine the evolution of the ray until it intercepts a boundary or the computation is truncated.

Once a ray path between two points in the medium is found, the measurable time delay for a signal traversing the ray path is determined by the path integral

$$\tau_s = \frac{1}{c_0} \int n ds. \tag{2.27}$$

The distance along the path, which defines the true range, is

$$r_s = \int ds. \tag{2.28}$$

Whereas τ_s is a measurable signal delay, r_s must be inferred. Determining r_s is tantamount to identifying the spatial position or a family of possible spatial positions that connect the source with points that keep the path integral constant.

2.2   NUMERICAL SIMULATIONS

In the remainder of this chapter only the two-dimensional form of (2.21) will be used. The three-dimensional form will be explored in Chapters 3 and 4, but the two-dimensional form simplifies the graphical representation of results that illustrate the range of phenomena supported by the FPE. In Chapter 6 scattering boundaries will be introduced. Anticipating that extension, $\psi(x, z)$ will represent either the horizontal component of the electric field (horizontal polarization) or the horizontal component of the magnetic field (vertical polarization). In a two-dimensional medium the electric and magnetic field components lie within or transverse to the xz plane. The FPE can be written as

$$\frac{\partial \psi(x, z)}{\partial x} = ik\Theta\psi(x, z) + ik\delta n(x, z)\psi(x, z). \tag{2.29}$$

---

[6]The ray equation can be derived directly from (2.24)[30, Section 5.2].

The formal solution with $\delta n(x, z) = 0$ is

$$\psi(x, z) = \int \widehat{\psi}(x_n; \kappa_z) \exp\left\{ikg(\kappa_z)(x - x_n)\right\} \exp\{i\kappa_z z\} \frac{d\kappa_z}{2\pi}, \qquad (2.30)$$

where

$$\widehat{\psi}(x_n; \kappa_z) = \int \psi(x_n, z) \exp\{-i\kappa_z z\} dz. \qquad (2.31)$$

To implement a split-step solution to (2.29), the propagation medium is divided into slabs of thickness Δx_n, which may vary. Under the assumption that the amplitude change over the slab is small, the phase perturbation, $\exp\left\{ik\Delta x_n \delta n(x_n, z)\right\}$, is applied at the initiation of each layer. For now, assume that $\delta n(x_n, z)$ is specified by a deterministic functional form. Assuming the wave field $\psi(x_n, z)$ at the start of slab n is known, an application of (2.30) with $x = \Delta x_n$ will advance the field to the start of the next slab. The split-step recursion is defined by the following three operations, the second two of which are forward and inverse discrete Fourier transforms (DFTs):

$$\psi(x_n, m\Delta z) = \psi(x_{n-1}, m\Delta z) \exp\left\{ik\delta n(x_n, m\Delta z)\Delta x_n\right\} \qquad (2.32)$$

$$\widehat{\psi}(x_n; l\Delta\kappa_z) = \sum_{m=0}^{N-1} \psi(x_n, m\Delta z) \exp\{-2\pi i lm/N\} \qquad (2.33)$$

$$\psi(x_{n+1}, m\Delta z) = \frac{1}{N} \sum_{l=0}^{N-1} \widehat{\psi}(x_n; l\Delta\kappa_z) P_l^{\Delta x_n} \exp\{2\pi i lm/N\} \qquad (2.34)$$

where

$$P_l = \exp\{ikg(\kappa_z(l\Delta\kappa_z)\}. \qquad (2.35)$$

The spatial frequency vector in DFT order is

$$\kappa_z(l\Delta\kappa_z) = [0, 1, \cdots, N/2 + 1, -N/2, -N/2 + 1, \cdots, -1] \Delta\kappa_z.$$

The recursion is initiated by specifying the starting field $\psi(x_0, m\Delta z)$, which can be thought of as an equivalent aperture field distribution. The only condition on the step size is that the amplitude change over the layer be small. The parameters that must be defined are the initial field, the sampling intervals Δz and Δx_n, the DFT size N, and the number of slabs. The wavelength sampling is determined by the DFT relation $2\pi/N = \Delta z \Delta \kappa_z$. For a vertical extent Z, $dz = Z/N$, and $d\kappa_z = 2\pi/(Ndz)$, which implies that $-\pi/dz <= \kappa_z < \pi/dz$. It follows that

$$\sqrt{1 - (\max \kappa_z/k)^2} = \sqrt{1 - (\lambda/(2dz))^2}. \qquad (2.36)$$

The vertical support required for a specific problem is dictated by the maximum beam extent and the propagation distance. Sampling is dictated by embedded structure and the severity of the disturbance.

Regarding the frequency choice and the propagation distances for the examples, a scalar refractive index is applicable to ionospheric propagation only for frequencies above 30 MHz, which is formally designated as the end of the high-frequency (HF) band. The 10^9 Hz (1 GHz) frequency chosen for the first examples to be presented below allowed for rapid computation over a propagation space subtending 5 km vertically and a maximum propagation distance of 150 km. The diffraction integral was evaluated using 8192 point DFTs. The number of slabs steps was 512 or 1024. The outputs at each slab are saved with supporting variables sufficient to reproduce the displays and to perform additional analyses or continue the forward propagation.

2.2.1 Beam Propagation

The first example illustrates the most basic application of the forward propagation code, namely the propagation of a beam in free space. Using (1.21), the following definition characterizes the modulation factor for such a freely propagating beam:

$$F(x, z) = \frac{\sqrt{-2\pi i/k}}{\max \left| g(\kappa_r)\widehat{\psi}(0; \kappa_r) \right|} \sqrt{r}\psi(x, z) \exp\{-ikr\}. \qquad (2.37)$$

As discussed at the end of Chapter 1, the normalization to the beam maximum avoids division by small numbers at the extrema of the field. Figure 2.1 is a grayscale intensity display of (2.37) at each slab. The field at the first slab is a raised cosine subtending 50 m. The cosine taper is easily generated and produces a representative beam structure. The computation was performed using a direct implementation of the split-step recursion. No random refractive index variation was employed in this example. The normalized field is presented in dB units.[7]

As discussed in Section 1.2.4, a beam represents a physical excitation source. Unit intensity on the primary axis of the scaled beam (0 dB) indicates no deviation from the expected on-axis intensity. Errors occur mainly because of aliased energy that enters the computation grid at the finite boundaries. This can be seen more clearly in Figure 2.2, which shows a comparison of the far field computed from (1.21) to that computed using the spectrum of the source field. Aliasing errors increase both with distance from the source and with distance from the axis of the beam, which encourages the use of narrow beams for analytic computation. In principal one can eliminate boundary reflections by imposing transparent boundary conditions. However, the procedures are intricate and can be computationally demanding. For the analysis in this book periodic boundary conditions and, if necessary, edge tapering are sufficient.

---

[7]Decibel intensity is defined as $10 \log 10(I)$, where $I = |v|^2$ and v is the complex field amplitude. The reference level is the far-field on-axis intensity.

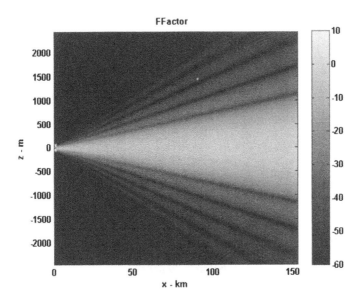

Figure 2.1   Beam propagation in homogeneous medium.

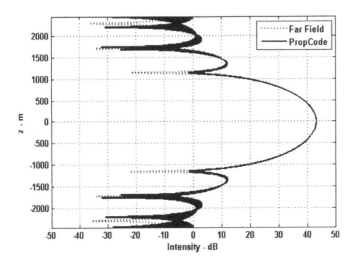

Figure 2.2   Comparison of FPE and far-field approximation.   Aliased energy distorts the beam sidelobes near the computation grid boundary.

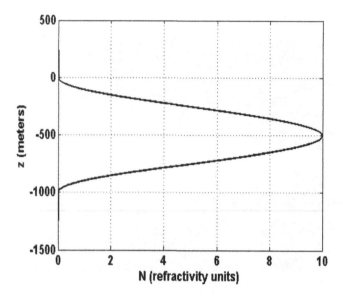

Figure 2.3   Refractive index profile for uniform horizontal layer.

2.2.2   Refraction

In the second example the beam is launched at a small downward propagation angle (-0.3 degrees) so it will pass through a layer of enhanced refractivity. Refractive index changes are generally reported in refractivity units

$$N = (n - 1)10^6. \qquad (2.38)$$

It follows that $\delta n = N10^{-6}$. Figure 2.3 shows the refractivity profile applied at each slab to simulate the effects of a uniform layer parallel to the propagation reference axis. Figure 2.4 shows the forward propagation of the beam in the presence of the enhanced layer. The layer bends the beam downward, but due to the spread of energy in the beam and the varying refractivity, the beam shape is slightly distorted as well. If one were to observe the beam in a plane at ~ 150 km from the source, the displacement of the peak from its expected location would suggest that a refracting layer is acting on the beam.

Additional insight can be gleaned from the evolution of the spatial wavenumber spectrum, which is an intermediate step in the forward propagation computation. Figure 2.5 shows the evolving spectrum plotted against the grazing angle. The spectrum at $x = l_p$ is the spectrum of the raised cosine, which is shifted to -0.3 degrees because of the phase ramp applied to the aperture distribution. In the absence of the layer, the spectrum would remain invariant. Where the main beam energy intercepts the layer at ˜100 km, the principal

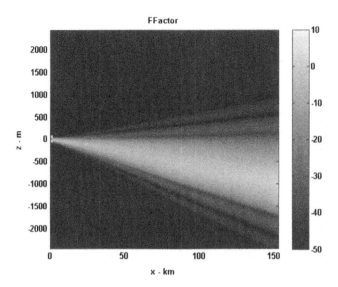

Figure 2.4 Downward propagation of beam through refracting layer.

direction (as indicated by the spectral peak) is refracted downward. As the beam propagates beyond the layer, the spectrum develops a narrow secondary peak at -0.3 degrees although the overall propagation direction is shifted to -0.4 degrees. In Section 2.2.3 ray paths will be overlaid to show this more clearly.

The final refraction example illustrates very strong refraction. This is achieved by launching the beam horizontally within a enhanced refractive index layer. To force the desired focusing effect within the computation window, the peak refractive index (shown previously in Figure 2.3) was increased from 10 to 40 refractivity units. The layer thickness was decreased to keep the incremental intensity changes small. Figure 2.6 shows the propagation result. As the beam expands from its point of origin, refraction redirects the energy back toward the beam axis creating a focus and virtual source at ~ 110 km. Some energy is not redirected, but significant on-axis signal strength is sustained well beyond the normal free-space range. There is no naturally occurring phenomenon that would produce the effect, but it illustrates the behavior of energy confined to a structure such as an optical fiber. Dramatic self-focusing effects under strong-scatter conditions where chance configurations of refractive-index variations produce similar focusing phenomena will be demonstrated in Chapter 3. Figure 2.7 shows the evolution of the spatial wavenumber spectrum. One can see clearly continual redirection of the plane wave decomposition while the energy remains well within the narrow scatter

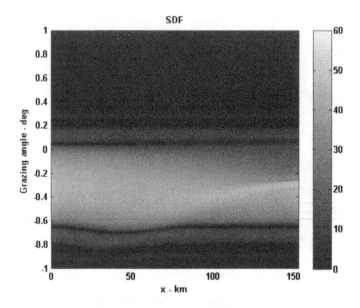

Figure 2.5   Progression of wave field spectrum for propagation through refracting layer.

range. Ray optics methods would produce a similar picture, but more often ray direction alone suffices for geometric constructs.

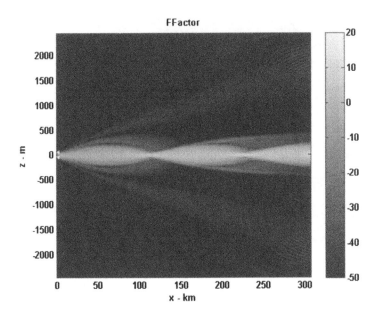

Figure 2.6   Example of beam propagation within a strongly refracting layer.

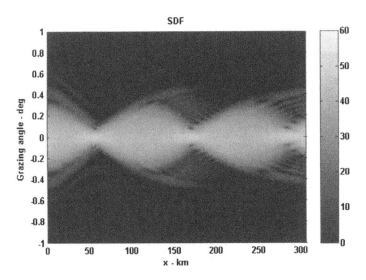

Figure 2.7   Evolution of spatial wavenumber spectrum of field propagating in strongly refracting layer.

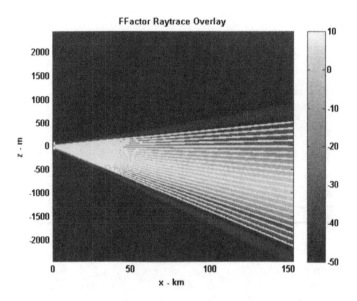

Figure 2.8   Ray bundle overlaid on FPE simulation from Figure 2.5.

2.2.3   FPE and Ray Optics

Ray optics was introduced following the parabolic approximation to the wave equation. Ray optics methods are used extensively for large-scale refraction computations and occasionally for scintillation simulations. Ray concepts have also been used to develop phenomenological scintillation models [31]. A direct solver for (2.26) has been implemented for computations in extended refracting media. The input to the calculation is a functional representation of the refractive index variation and the starting directions for a bundle of rays. Figure 2.8 shows an overlay of a ray bundle on the forward propagation simulation shown in Figure 2.4. The rays, which emanate from a point, follow the trajectory of the central refracted beam as one expects. The fine structure of the beam intensity is not included in the ray trace computation. Because most practical computations are well within the range of the parabolic approximation, which is a necessary condition for ray tracing, the two methods are complementary. Ray tracing is often applied to a smooth background representation of the propagation medium. Structure is then incorporated as an incremental perturbation along the ray path.

2.2.4 Scintillation

The examples presented thus far have emphasized beam propagation with analytic background departures from strict homogeneity. Beamforming antennas are used to achieve the sensitivity required for detection. Although the known antenna gain patterns can be incorporated in simulations, source-to-destination computation is demanding and unnecessary for many applications. The far-field approximation shows that measured fields are locally planar. Thus, initiating a scintillation calculation with a plane provides a direct measure of the field modulation. The means of incorporating spreading losses and corrections for wavefront curvature will be described in Chapters 4 and 5.

A plane wave propagates without amplitude change in a homogeneous medium. The discrete representation of a plane wave to which a sampled propagation operator is applied also propagates without amplitude change. The resulting pseudo-conservation occurs because energy that formally crosses a computation grid boundary reenters the grid at the opposite boundary. However, a point can be reached where the energy leaving and reentering a computation grid fundamentally alters the field being realized. A rigorous test for aliasing is affected by increasing the size of the computation grid and repeating the computation. The downsampled result should agree with the initial result if the computation grid is adequate. However, monitoring the energy in the angular spectrum at the largest resolved wavelength is generally adequate.

A plane wave is initiated in the two-dimensional implementation of the split-step algorithm with a starting field of the form $\exp\{2\pi i k \sin(\theta) n\Delta z\}$. The angle θ determines the propagation direction. Complex field modulation develops as the wave encounters structure mapped initially onto the wave field phase. Because the process is cumulative, scintillation structure will continue to build in intensity until a steady or saturated state is achieved. How saturation is reached and how the transition from weak scatter to saturation can be predicted is discussed in Chapter 3.

The means of generating refractive index realizations that can produce representative in situ structure will be discussed in detail in Chapters 3 and 4. By way of introduction, consider the weighted Fourier transformation of a set of uncorrelated unit-variance, zero-mean, Hermitian Gaussian random variates ζ_m :

$$\delta n(n\Delta z) = \sum_{m=0}^{N-1} \sqrt{\Phi_{\delta n}(m_p(m)\,\Delta\kappa_z)\Delta\kappa_z/\,(2\pi)}\zeta_m \exp\{2\pi i n m/N\}. \quad (2.39)$$

The notation $m_p(m)$ means DFT order

$$m_p([0,1,\cdots,N]) = [0,1,\cdots,N/2,-N/2+1,\cdots-2,-1]. \quad (2.40)$$

Hermitian conjugate symmetry makes $\delta n(n\Delta z)$ purely real. The non-negative spectral density function (SDF) $\Phi_{\delta n}(m_p(m)\,\Delta\kappa_z)$ is symmetric about zero wavenumber. The DFT construction assumes the periodicity $\delta n(n\Delta z) = \delta n((m+lN)\Delta z))$ for any l. The Hermitian property translates to purely real ζ_0 and $\zeta_{N/2}$ with $\zeta_{N-m} = -\zeta_m^*$ for $m = 1, 2, \cdots, N/2-1$. Uncorrelated variates with unit-variance have the formal property

$$\langle \zeta_m \zeta_{m'}^* \rangle = \delta(m - m'). \tag{2.41}$$

The angle brackets denote an ensemble average.[8]   Note that the DFT of $\delta n(n\Delta z)$ is $\sqrt{\Phi_{\delta n}(m_p(m)\,\Delta\kappa_z)\Delta\kappa_z/(2\pi)}\zeta_m$. It follows from (2.41) that

$$\left\langle \left| \sqrt{\Phi_{\delta n}(m_p(m)\,\Delta\kappa_z)\Delta\kappa_z/(2\pi)}\zeta_m \right|^2 \right\rangle = \Phi_{\delta n}(m_p(m)\,\Delta\kappa_z)\Delta\kappa_z/(2\pi), \tag{2.42}$$

and that

$$\langle \delta n(n\Delta z)^2 \rangle = \sum_{m=0}^{N-1} \Phi_{\delta n}(m\Delta\kappa_z)\Delta\kappa_z/(2\pi). \tag{2.43}$$

Thus, (2.39) creates a realization of a random process with SDF $\Phi_{\delta n}(m\Delta\kappa_z)$. The properties of stochastic processes that underlie these manipulations will be reviewed in Chapter 3.

The simplest SDF that is supported by both theory and measurement is the power-law form

$$\Phi_{\delta n}(q) = Tq^{-p}, \tag{2.44}$$

where T is the turbulent strength and p is the power-law index. For the examples presented here, the power-law range extends from the smallest resolved wavenumber $q = \Delta\kappa_z$ to the Nyquist frequency $q = N\Delta\kappa_z/2$. This means that the variance of the realization will increase as N is increased. As will be discussed in detail in Chapter 3, this behavior is representative of real-world measurements. The important point to note is that the turbulent strength and the power-law index are robust observables. This places particular emphasis on SDF measurements for diagnostic purposes. For most radio observations, the structured region is confined to a subregion of the propagation space.

The first scintillation example uses a single slab with the phase profile shown in Figure 2.9. Figure 2.10 shows the evolving field intensity. The incident field intensity is normalized to unity. The structured layers are marked with asterisks on ordinate axis. As already noted, with plane-wave excitation there is no variation in the average intensity. The evolving intensity scintillation has as a filamentary structure that suggests local focusing by random lens-like structures.

---

[8]An ensemble average reproduces a value that represents the limiting form of an average over a very large number of realizations of the process.

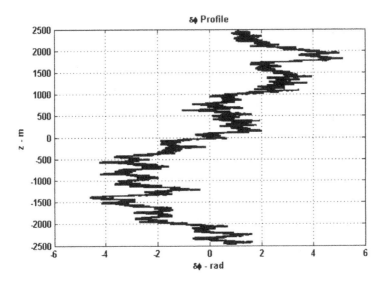

Figure 2.9   Power-law refractive index structure profile scaled to radians.

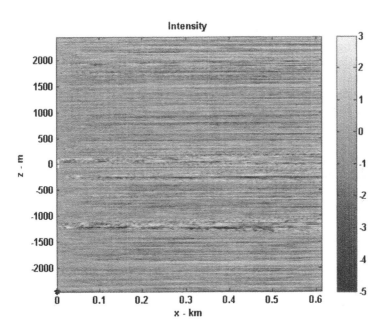

Figure 2.10   Evolving intensity field from single phase screen with a shallow power-law phase structure.

A standard measure of the level of scintillation intensity is the scintillation index (SI), which is defined as

$$SI = \frac{\sqrt{\langle I^2 \rangle - \langle I \rangle^2}}{\langle I \rangle}. \tag{2.45}$$

Figure 2.11 shows an estimate of the evolving SI index obtained from the field at each slab. How the intensity structure evolves will be discussed in detail in Chapter 3. Here the relation between an *equivalent* phase screen and an extended region will be explored. Figure 2.12 shows the complex intensity and the complex field at the maxim distance 602.78 m. Figure 2.13 shows the intensity and phase structure of the field after propagating twice that distance in a continuous medium with the same integrated structure. That is, the initial layer is equivalent to concentrating the distributed structure in a single layer placed at the center of the disturbed region. The similarity of the structure verifies the equivalence predicted by the weak-scatter theory. Figure 2.14, which can be compared to Figure 2.10, and Figure 2.15, which can be compared to Figure 2.11, show the details of the evolution of the structure with distance. It can be seen that the slower evolution of the field in the extended medium is compensated by the extra distance over which the propagation occurs. This phase-screen equivalence greatly simplifies analytic computations. It was recognized early in the development of the theory, and it will be exploited fully in Chapter 3.

The evolving structure of the complex field needs further elaboration. The forward translation of refractive index structure to path-integrated phase is unambiguous. However, the phase of a complex field is 2π ambiguous. In a well-sampled one-dimensional phase reconstruction using the arctangent function phase discontinuities are easily detected and removed as long as the signal strength is adequate. The complex field structure shown in the lower frames of Figures 2.12 and 2.13 shows large-scale variation through many 2π cycles over the measurement window. The amplitude variation maps out this phase meander as an annulus in the complex plane. This large-scale phase structure is a consequence of the small angle at which the plane wave propagates. As such, it is purely deterministic and could be removed, but it was included as a reminder that in with real-world measurements, establishing a phase reference involves a choice of scale. The fact that the amplitude variation shown in the upper frames of Figures 2.12 and 2.13 is more homogeneous is a consequence of the diffraction process as will be described in more detail in Chapter 4.

The final scintillation example shows the same computation with the total phase structure introduced incrementally from layer to layer rather than being concentrated in a single layer. Figure 2.14, which can be compared to Figure 2.10, shows the result. The corresponding scintillation index is shown in Figure 2.15. In a distributed medium the scintillation builds up more gradually, but ultimately achieves a similar structure to the single phase screen.

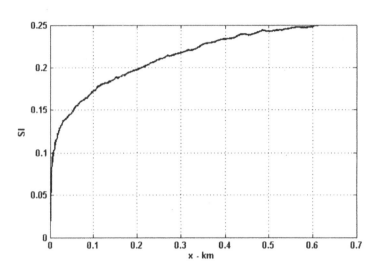

Figure 2.11 Scintillation index versus distance for evolving field structure shown in Figure 2.10.

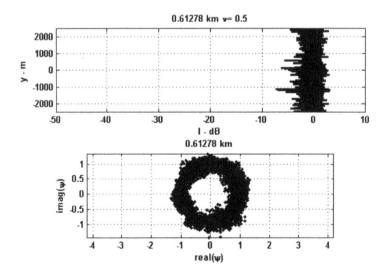

Figure 2.12 Intensity and phase structure at approximately one kilometer from the initiating phase screen.

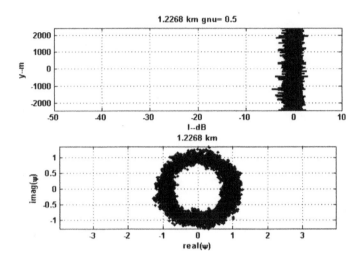

Figure 2.13   Fully developed intensity and phase structure at a large distance from the initiating phase screen.

The two examples are similar but not strictly equivalent.   With regard to statistical similarity, the results compare surprisingly well.   To conclude the scintillation introduction, Figure 2.16 shows the SDF in the evolving structure.   The evolving SDF shows persistent frequency enhancements suggestive of the filamentary structure seen in Figures 2.10 and 2.14.   However, the significant energy content is carried by well-resolved spatial wavenumbers.

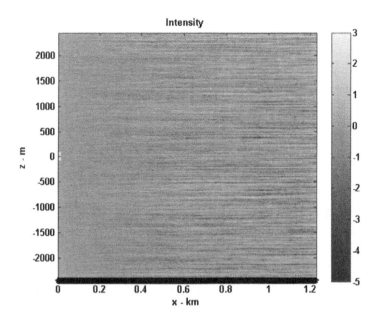

Figure 2.14   Evolving intensity structure for extended propagation disturbance of the same total RMS phase as imparted by the single phase screen.

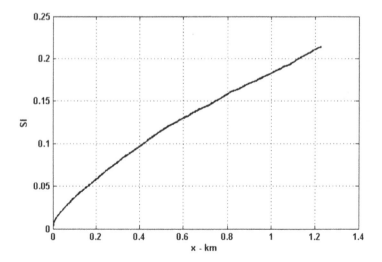

Figure 2.15   Evolving SI index for distributed layer.

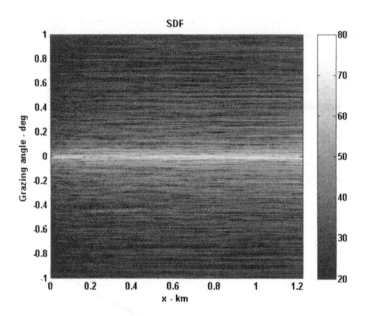

Figure 2.16 Evolving spectral density function for distributed structure.

CHAPTER 3

THE STATISTICAL THEORY OF SCINTILLATION

As far as the laws of mathematics refer to reality, they are not certain; and as far as they are certain, they do not refer to reality.

—*Albert Einstein*

This chapter develops the statistical theory of scintillation as an analytic framework for computing field moments from a statistical characterization of the random medium and the propagation geometry. A tractable formulation is made possible by imposing a high degree of regularity on the statistical structure of the propagation medium. The formal theoretical result is a system of differential equations that can be solved, in principle, for any desired complex field moment. Each moment equation is similar in structure to the forward propagation equation (FPE) in that it is comprised of propagation components and additive media-interaction components. Constructing this system of equations was a major accomplishment in the development of scintillation theory. The work of Russian theoreticians disseminated through the translation of V. I. Tartarski's book [32], *The Effects of the Turbulent Atmosphere on Wave Propagation*, stimulated the theoretical effort in the early 1970s. These contributions are summarized in the review by Barabanenkov,

The Theory of Scintillation with Applications in Remote Sensing, by Charles L. Rino **45**
Copyright © 2011 Institute of Electrical and Electronics Engineers

et al. [33]. Most of the theoretical work since the early 1970s has been devoted to solving and interpreting the moment equations. A complete mathematical derivation of the moment equations was published in 1974 by L. C. Lee [34].

The family of differential equations that define the field moments of all orders will be developed in this chapter as a formal extension of the FPE. Any such development will encounter the problem of calculating expectations that involve products of functions of field observables with products of functions of the media-interaction terms that impart the structure to the field observables. The mathematical tools tailored to solve such problems are unfamiliar to many researchers involved in remote sensing, but the critical results can be comprehended and used effectively with the constructs developed in this chapter. References are included for readers interested in more complete mathematical details.

The chapter starts with an overview of how structure is imparted to the material properties of representative propagation media. Induced structure is typically manifest as random variations of the thermodynamic state variables of the gas or fluid that comprises the propagation medium. The theory of stochastic processes provides the mathematical framework to characterize these random fields. The theory of fluid turbulence motivates using a power-law spectral density function (SDF) as a reference structure model. Kolmogorov turbulence is characterized by a three-dimensional radial wavenumber power-law SDF of the form $q^{-11/3}$. This model is strictly applicable only to neutral atmospheric turbulence, but the general power-law model is appealing for scintillation applications because it leads to a compact parameterization of the scintillation phenomenon that agrees well with observations. Even so, solving moment equations is a formidable task with many possible approaches. To capture the essential elements of the theory, only the results that can be reconstructed in a compact form are presented. This constrains the analytic results to the single phase screen model, but an attempt has been made to describe analysis techniques that have pioneered our broader understanding of scintillation phenomena.

To characterize scintillation structure, intensity correlations derived from fourth-order moments are essential. To calculate waveform distortion and coherence loss, two-frequency and mutual coherence functions (second-order moments) are essential. However, a complete statistical description of scintillation phenomena must include first-order statistics as well. Because of the complexity of this problem, purely phenomenological models have been most successful. Thus, a review of the phenomenological theory of first-order scintillation statistics is presented in this chapter as a complement to the statistical theory. The chapter concludes with numerical simulations that support and extend the theoretical results, particularly as they apply to intensity statistics. (See Tables 3.1 and 3.2 for symbols and abbreviations.)

Table 3.1 Chapter 3 Symbols

Symbol	Definition
$n(\mathbf{r};t) = c/c_0 = \sqrt{\mu\epsilon}/\sqrt{\mu_0\epsilon_0}$	Refractive index
n_0, $n_d(\mathbf{r}(t))$, $n_t(\mathbf{r}(t);t)$, $\delta n(\mathbf{r}(t);t)$	Constant, Deterministic, Slow Varying, Random
N	Refractivity (3.2)
ω_p	Plasma frequency (3.6)
$\langle\cdot\rangle$	Ensemble average
$\Gamma_{nm}(x;\zeta_1,\cdots\zeta_N;\xi_1,\cdots\xi_M)$	Complex field moment of order NM (3.43)
$\Phi(\mathbf{K})$	Spectral density function of $\delta n(\mathbf{r})$ (3.13)
$\Re(y)$	Unit-normalized isotropic ACF (3.17)
$Q(q)$	Unit-normalized isotropic SDF (3.18)
$R_{\delta n}(y)$, $\Phi_{\delta n}(q)$	ACF, SDF of $\delta n(\mathbf{r})$
$D_{\delta n}(y)$	Structure function (3.26)
$q_S = 2\pi/l_S$, $q_L = 2\pi/l_L$	Small-scale, large-scale wavenumbers ($l_L >> l_S$)
ν	Power-law index parameter $Q(q) \sim q^{-(2\nu+1)}$
$K_\eta(x)$	Modified Bessel function fractional order η
C_s	Power-law turbulent strength $Q(q) \simeq C_s q^{-(2\nu+1)}$
C_n^2	Power-law structure constant $D(y) \simeq C_n^2 y^{2\nu-2}$
$\delta\phi(\varsigma)$, $\delta\bar{n}(\varsigma)$	Path-integrated phase, refractive index
$R_{\delta\phi}(y)$, $\Phi_{\delta\phi}(q)$, $D_{\delta\phi}(y)$	ACF, SDF, and SFN of path-integrated phase
$\varkappa$	Phase ACF normalization factor (3.36)
l_P	Layer thickness
$C_p = k^2 l_p C_s$	Phase turbulent strength
$\rho_F = \sqrt{x/k}$	Fresnel radius
$\Phi_I(\kappa)$	Intensity SDF
$p_A(a)$ or $p(A)$	PDF of random variable A or its value

Table 3.2   Chapter 3 Abbreviations

Abbreviation	Definition
ACF	Autocorrelation function
SFN	Structure function
PDF	Probability density function
TEC	Total electron content

3.1   BACKGROUND

The statistical theory of scintillation builds on mathematical statistics, particularly the theory of stochastic processes, which includes spectral representations that have already been introduced. It is also important to understand the sources of in situ structure. The background material presented here reviews these topics to establish the broader framework that underlies the development of the statistical theory of scintillation.

3.1.1   Structure Sources

Scintillation originates from structure in the propagation medium. Small-scale irregular structure is the primary concern of the statistical theory, but structure development involves a hierarchy of scale-dependent processes. It is important to consider at the outset the complete propagation environment. The following refractive-index formulation identifies three distinct contributions to spatial and temporal variations of the refractive index:

$$n(\mathbf{r}; t) = 1 + n_d(\mathbf{r}(t)) + n_t(\mathbf{r}(t); t) + \delta n(\mathbf{r}(t); t). \tag{3.1}$$

The deterministic component $n_d(\mathbf{r}(t))$ accommodates variations that have a known average structure with negligible temporal variation over the duration of a typical observation. With a vacuum as a reference, $n_d(\mathbf{r}(t))$ necessarily includes a constant representing a strictly homogeneous background. Some effects of deterministic profiles (layers of enhanced refractive index) were illustrated in the examples presented in Chapter 2. The component $n_t(\mathbf{r}(t); t)$ accommodates trend-like variations that are driven by processes that cannot be characterized by statistically homogeneous measures. Large-scale gradients where instabilities initiate structure development are examples of the intermediate structure component. The residual structure represented by $\delta n(\mathbf{r}(t); t)$ accommodates the purely random component, which is the main topic of this chapter. Although the refractive-index structure breakdown is well-defined as a construct, it is not possible to separate the components unambiguously from realizations of $n(\mathbf{r}; t)$ alone.

The dominant component of the Earth's atmosphere is a neutral gas and vapor mixture. Because changes in the refractive index on the order of 10^{-6} have measurable effects at microwave and optical frequencies, it is convenient to employ refractivity units

$$N = (n - 1)10^6. \tag{3.2}$$

The following formula, derived by Smith and Weintraub [35], relates atmospheric refractivity to atmospheric pressure, (p in millibars) temperature (T in degrees Kelvin), and water vapor pressure (e in millibars):

$$N = \frac{77.6}{T}(p + \frac{4810e}{T}). \tag{3.3}$$

In an ideal gravity-supported atmosphere, pressure decreases exponentially with increasing height. The average atmospheric refractivity would decrease exponentially with increasing height only under conditions of uniform temperature and humidity. Thus, in 1959 the Central Radio Propagation Laboratory (CRPL) developed an average reference atmospheric refractivity model [36], which is still used to estimate the expected refraction of electromagnetic waves in the Earth's atmosphere. Figure 3.1 is a plot of the CRPL reference refractive-index height profile. The dashed curve is the simplified exponential approximation

$$N = N_s \exp\{-0.139h_{km}\}, \tag{3.4}$$

with $N_s = 303$, which can be taken as the constant background contribution. The independent variable h_{km} is the height above the surface measured in kilometers. The reference atmosphere is an example of the $n_d(\mathbf{r})$ component of (3.1). Atmospheric meteorological conditions cause significant local departures from the mean background. These structures admit diurnal, seasonal, and geographic variations [37]. It is this type of variation that would be accommodated by the $n_t(\mathbf{r})$ component of (3.1).

How turbulent atmospheric structure develops and how it is transmitted to the refractive index field is an involved subject. Fluid turbulence is manifest mainly in the velocity field, and the same is true of the atmosphere. However, there are profound differences between an incompressible fluid and a gas, but essential elements of turbulence theory apply. Scalar variables such as temperature and pressure respond to the turbulent flow field as *passive scalars*, which means the scalar fields acquire the same turbulent characteristics as the radial velocity field. Sensitive temperature probes are commonly used to measure turbulent structure. Small temperature fluctuations about a steady mean are manifest as refractive index variations. Scintillation measurements provide a path-integrated measure of the turbulent structure. Since the

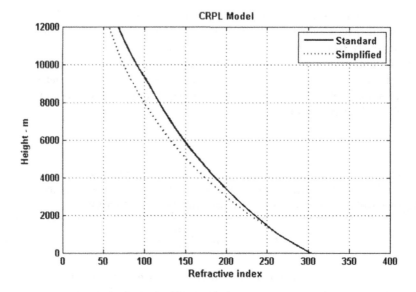

Figure 3.1   Model atmospheric refractivity profiles.

details of how that structure develops do not impact the theory of scintillation, this topic is left for the reader's exploration.

The Earth's atmosphere transitions to the ionosphere above approximately 90 km. The Earth's ionosphere is a plasma, which contains free charged particles that respond almost freely to electromagnetic forces. The response of plasmas to electromagnetic forces remains an active subject of research that was stimulated in its early development by radio propagation observations. The treatment of this subject by Yeh and Liu [9, Chapter 3] is particularly well-suited to remote sensing and scintillation applications. The book by K. G. Budden captures the art of early radio science [38].

At frequencies below 30 MHz, the ionosphere supports four polarization-dependent propagation modes. The formal constitutive relationship that captures this behavior involves a tensor permittivity that replaces the scalar permittivity model used throughout the development in this book. The applications that are of most practical concern for scintillation theory, however, use higher frequencies where the scalar cold-plasma approximation

$$n_p = \sqrt{1 - (\omega_p/\omega)^2} \tag{3.5}$$

is valid. The plasma frequency, ω_p, is defined by the relation

$$
\begin{aligned}
\omega_p^2 &= N_e e^2 / (m_e \varepsilon_0) \\
&= 4\pi r_e N_e c^2
\end{aligned}
\tag{3.6}
$$

where N_e is the electron density, e is the electron charge, m_e is the electron mass, and r_e ($2.8197402894 \times 10^{-15}$ m) is the classical electron radius. Since the refractive index is purely imaginary below the plasma frequency, propagating EM fields are effectively excluded where the radio frequency is below the plasma frequency. An electromagnetic wave propagating into such a *cut-off* region will be reflected. An ionospheric sounder measures this reflected wave as a function of frequency. The resulting ionogram can be inverted to determine the electron density.

Even so, because of the inverse frequency dependence in (3.5), the ionosphere becomes more sensitive to small-scale structure as frequency decreases, which is the opposite of the frequency dependence of atmospheric scintillation. It follows from (3.5) that

$$
\begin{aligned}
\delta n_p &= -4\pi r_e \delta N_e c^2 / \omega^2 \\
&= -4\pi r_e \delta N_e / k^2,
\end{aligned}
\tag{3.7}
$$

whereby $\delta\phi = k\delta n_p \propto 1/f$. Replacing $k\delta n$ with $-2r_e\lambda\delta N_e$ introduces an explicit dependence on the electron density perturbation. Although electron density fluctuations map directly to refractive index fluctuations, the frequency dispersive effect profoundly alters the occurrence and structure of scintillation in ionized media.

Dual-frequency phase-difference measurements can be used to estimate the integrated total electron content (TEC) [39]. An example will be presented in Chapter 5. The interest in TEC stems from the fact that ionospheric refractivity profiles are driven by electron content structure, which responds directly to electromagnetic forces. Consequently the components of ionospheric structure are more variable than their atmospheric counterparts. Moreover, a substantial component of ionospheric structure populates the regime between deterministic and statistically homogeneous. Nickisch developed a stochastic integrated electron content model for the polar region to characterize the large-scale content that is not directly manifest as intensity scintillation [40]. This is an example of an explicit characterization of the intermediate scale structure that dominates total electron content measurement.

In spite of these dramatic differences in structure sources and scale, the relations (3.3) and (3.7) characterize refractive-index structure supported by a common parametric model that can be used to interpret atmospheric scintillation, ionospheric scintillation, and galactic scintillation. The formal structure of the FPE can be applied to sound propagation as well. Sound has its own unique characteristics related to transmission through solids and liquids. Sound propagation in the ocean shares many of the refraction characteristics of high-frequency waves propagating in the ionosphere.

3.1.1.1 Temporal Variation Scintillation measurements are recorded most often as time series. The temporal variation generated by a combination of source motion, receiver motion, irregularity drift, and structure evolution.

Under conditions where source and receiver motions are known, one can isolate a critical component of time variation. Temporally varying positions as observed in the propagation reference frame can be approximated as

$$\mathbf{r}(t) = \mathbf{r}(t_0) + (t - t_0)\mathbf{v} + \cdots, \tag{3.8}$$

with higher-order terms added as a particular scenario dictates. Temporal variation of the structure itself is usually incorporated as a multiplicative correction to specific statistical measures. Time variation has been neglected in the development of propagation theory thus far because the structures of primary interest are effectively stationary over the time interval that wave fields propagate through the disturbance. Under these conditions, time variation can be modeled as a velocity-dependent displacement. Measurements from spatially displaced receivers can be used to test the hypothesis of stationary structure. Satellite scintillation observations, where the geometric contribution to time variability dominates, will be developed in detail in Chapter 4.

3.1.2   Stochastic Processes

The theory of stochastic processes provides the mathematical framework for generating statistical measures that characterize random structure. A stochastic process is defined formally by an ensemble of field realizations. The relevant fields here can be defined mathematically as complex functions of dependent variables that include space, time, and frequency. Formal ensemble averages of any operation involving field observables provide deterministic measures of average field attributes. The realizations that are used most often involve fields with unbounded support. Time series and spatially evolving fields are examples. Measure theory supplies the theoretical framework. The textbook by Athanasios Papoulis is still one the best intermediate-level treatments of the essential material [41].

Ergodicity is an important property of stochastic processes that ensures that a single realization of the process embodies structure that reflects the underlying ensemble averages. For example, the summation

$$\langle \delta n(\mathbf{r})\delta n(\mathbf{r}') \rangle \sim \frac{1}{N} \sum_{n=1}^{N} \delta n(\mathbf{r}_n)\delta n(\mathbf{r}'_n) \tag{3.9}$$

approximates the spatial covariance or autocorrelation function (ACF) implied by the ensemble average denoted by angle brackets. The attributes of the approximation, for example, its mean and variance, depend on the statistical properties of the process and the environment in which the measurement is made.

Statistical homogeneity is the critical formal property that supports modeling of in situ structure and the generation of stable estimates of the statistical measures just described. In the development of propagation theory, homogeneity refers to spatial and temporal invariance of the fields. Statistical homogeneity refers to spatial and temporal invariance of statistical measures. Weak statistical homogeneity refers to spatial and temporal invariance of moments defined by pairs of space-time samples, as in (3.9). As discussed in Section 3.1.3, weak statistical homogeneity is a necessary condition for the existence of the SDF measure used extensively in the development.

The statistical characteristics of the random index of refraction variations are established by the hierarchy of field moments

$$\Gamma_{NM} = \langle P_{NM} \rangle, \tag{3.10}$$

where

$$P_{NM} = \prod_{n=1}^{N} \delta n(\mathbf{r}_n) \prod_{n=1}^{M} \delta n(\mathbf{r}'_n) \tag{3.11}$$

is a shorthand for displaced products of refractive index samples. These moments characterize the degree of coherence between the sample sets comprising the products that form the particular moment. The lowest-order moment $\Gamma_{11}(\mathbf{r}_1, \mathbf{r}'_1)$ has already been introduced as the ACF. The class of processes for which the autocorrelation function depends only on the difference vector $\Delta\mathbf{r} = \mathbf{r}'_1 - \mathbf{r}_1$ are of particular interest.

Processes whose statistical characteristics can be constructed in terms of a small number of moments are of particular interest. The Gaussian random process has a number of additional properties that make it particularly appealing as a model for the random structure component. For example, linear operations on Gaussian processes yield Gaussian processes. The central limit theorem, which states that summations of realizations from any well-behaved process converge to Gaussian processes, justifies its widespread use. If, for example, x is a relization of a Gaussian random process, the characteristic function takes the particularly simple form

$$\langle \exp\{ix\} \rangle = \exp\left\{i\langle x^2 \rangle / 2\right\}. \tag{3.12}$$

The fact that a Gaussian process is completely characterized by its second-order moments is also important. Hereafter, the refractive index structure will be assumed to be characterized by a three-dimensional Gaussian random process.

3.1.3 Spectral Representation

Spectral representations of stochastic processes are essential both to characterize the in situ structure and to develop robust statistical measures. Statistically homogeneous stochastic processes admit spectral representation of the form

$$\delta n(\mathbf{r}) = \iiint \exp\{i\mathbf{K} \cdot \mathbf{r}\}\sqrt{\Phi(\mathbf{K})}d\xi(\mathbf{K}), \tag{3.13}$$

where the variable $\mathbf{K}$ denotes spatial wavenumber, and $d\xi(\mathbf{K})$ has the formal orthogonal increments property

$$\langle d\xi(\mathbf{K})d\xi^*(\mathbf{K}')\rangle = \delta\left(\mathbf{K} - \mathbf{K}'\right)\frac{d\mathbf{K}}{(2\pi)^3}. \tag{3.14}$$

It follows that

$$\langle \delta n(\mathbf{r})\delta n(\mathbf{r}')\rangle = \iiint \exp\{i\mathbf{K} \cdot \mathbf{\Delta r}\}\Phi(\mathbf{K})\frac{d\mathbf{K}}{(2\pi)^3}, \tag{3.15}$$

which establishes the Fourier transform relation between the ACF, denoted

$$R_{\delta n}(\mathbf{\Delta r}) = \langle \delta n(\mathbf{r})\delta n(\mathbf{r}')\rangle, \tag{3.16}$$

and the SDF, denoted $\Phi(\mathbf{K})$. The relation is often referred to as the Wiener-Khinchin theorem.

3.1.4 Power-Law Spectral Models

A theoretical development could proceed using only the homogeneous statistics assumption and correlation functions. However, the results do not admit analytically tractable solutions.   Fortunately, the properties of power-law processes can be exploited to generate analytic results that characterize scintillation phenomena that are routinely observed.   Turbulence theory has established that the three-dimensional SDF of a passive scalar in the turbulent flow field can be characterized by the power-law form q^{-n} over the spatial wavenumber range $q_L \ll q \ll q_S$. The smallest wavenumber, q_L, is referred to as the outer scale, which represents the largest structure.   The largest wavenumber, q_S, is referred to as the inner scale, which represents the smallest structure.[9]   To translate this characterization into a form suited for analytic manipulation, Shkarofsky [42] proposed the following self-similar ACF and spectral SDF forms:

$$\Re(y) = \frac{\left(\sqrt{1 + (q_S y/2)^2}\right)^{\nu-1} K_{\nu-1}(2\frac{q_L}{q_S}\sqrt{1 + (q_S y/2)^2})}{K_{\nu-1}(2\frac{q_L}{q_S})} \tag{3.17}$$

$$Q(q) = \left(\frac{4\pi}{q_S q_L}\right)^{3/2}$$

$$\times \frac{\left(\sqrt{1 + (q/q_L)^2}\right)^{-(\nu+1/2)} K_{\nu+1/2}(2\frac{q_L}{q_S}\sqrt{1 + (q/q_L)^2})}{K_{\nu-1}(2\frac{q_L}{q_S})}. \tag{3.18}$$

---

[9]The relation between wavenumber and wavelength scale is $q = 2\pi/\lambda$.

The function $K_\nu(x)$ is the modified Bessel function. These isotropic functional forms are related by a one-dimensional Fourier transform

$$\Re(y) = \iiint Q(q) \exp\{i\mathbf{q} \cdot \mathbf{y}\} \frac{d\mathbf{q}}{(2\pi)^3}$$

$$= \int_0^\infty q^2 \frac{\sin qy}{qy} Q(q) \frac{dq}{2\pi^2}. \qquad (3.19)$$

The normalization is such that

$$\Re(0) = \int_0^\infty q^2 Q(q) \frac{dq}{2\pi^2} = 1. \qquad (3.20)$$

Note also that the $(2\pi)^3$ factor here resides in the spectral-domain integral. Complete ACF and SDF functions are constructed by introducing the refractive index variance $\langle \delta n^2 \rangle$:

$$R_{\delta n}(y) = \langle \delta n^2 \rangle \Re(y) \qquad (3.21)$$

$$\Phi_{\delta n}(q) = \langle \delta n^2 \rangle Q(q), \qquad (3.22)$$

whereby $R(0) = \langle \delta n^2 \rangle = \int_0^\infty q^2 Q(q) dq/2\pi^2$. If $q \gg q_L$, $q \ll q_S$, and $q_L/q_S \ll 1$, the small-argument form $K_\eta(2x) \approx 0.5\Gamma(|\eta|)x^{-|\eta|}$ shows that as long as $\nu > 1$, $\Phi_{\delta n}(q)$ has the power-law form

$$\Phi_{\delta n}(q) = \langle \delta n^2 \rangle Q(q)$$

$$\sim C_s \left(q_L^2 + q^2\right)^{-(\nu+1/2)}, \qquad (3.23)$$

where

$$C_s = \frac{\langle \delta n^2 \rangle (4\pi)^{3/2} \Gamma(\nu + 1/2)}{\Gamma(\nu - 1)q_L^{-2\nu+2}} \qquad (3.24)$$

is the turbulent strength. To summarize, for $q_L \ll q$, $\Phi_{\delta n}(q)$ has the pure power-law form $C_s q^{-(2\nu+1)}$, and the variance remains finite as long as $\nu > 1$.

Because the large-scale components that dominate the structure fluctuations are often trend-like rather than cleanly defined transitions imposed by an outer scale, the structure function

$$D_{\delta n}(y) = \langle \delta n^2 \rangle \langle [\delta n(r) - \delta n(r')]^2 \rangle$$

$$= \int_0^\infty q^2 \left[1 - \frac{\sin qy}{qy}\right] Q(q) \frac{dq}{2\pi^2}$$

provides a more robust characterization than the autocorrelation function. To demonstrate the weak dependence on scale measures, the Shkarofsky model and the equivalence

$$\langle \delta n^2 \rangle = \frac{\Gamma(\nu - 1)C_s}{(4\pi)^{3/2} \Gamma(\nu + 1/2)q_L^{2\nu-2}} \qquad (3.25)$$

are used to rewrite the structure function as

$$D_{\delta n}(y) = 2[R_{\delta n}(0) - R_{\delta n}(y)] = \frac{2\Gamma(\nu - 1)C_s}{(4\pi)^{3/2}\,\Gamma(\nu + 1/2)}$$

$$\times \left[\frac{1 - \left(\sqrt{1 + (q_S y/2)^2}\right)^{\nu-1} K_{\nu-1}(2\tfrac{q_L}{q_S}\sqrt{1 + (q_S y/2)^2})/K_{\nu-1}(2\tfrac{q_L}{q_S})}{q_L^{2\nu-2}}\right]$$

$$\sim \frac{2\Gamma(\nu - 1)C_s}{(4\pi)^{3/2}\,\Gamma(\nu + 1/2)}\left[\frac{1 - 2\,(q_L y/2)^{\nu-1}\,K_{\nu-1}(q_L y)/\Gamma(\nu - 1)}{q_L^{2\nu-2}}\right].$$

$$(3.26)$$

The numerator and the denominator approach zero as $q_L \rightarrow 0$, whereby L'Hôpital's rule can be used to evaluate the limit of the final term in square brackets as the ratio of derivatives. Using

$$\frac{d\,(q_L y/2)^{\nu-1}\,K_{\nu-1}(q_L y)}{dq_L} = -y\,(q_L y/2)^{\nu-1}\,K_{\nu-2}(q_L y),\qquad(3.27)$$

with $1 < \nu < 2$, the following scale-free relation is obtained:

$$\lim_{q_L \to 0} D(y) \sim \frac{2C_s}{(4\pi)^{3/2}\,\Gamma(\nu + 1/2)}\frac{\lim\limits_{q_L \to 0} 2y\,(q_L y/2)^{\nu-1}\,K_{\nu-2}(q_L y)}{\lim\limits_{q_L \to 0}(2\nu - 2)\,q_L^{2\nu-3}}$$

$$\sim \frac{2\Gamma(2 - \nu)C_s}{(4\pi)^{3/2}\,(\nu - 1)\,\Gamma(\nu + 1/2)2^{2\nu-2}}y^{2\nu-2}.\qquad(3.28)$$

This establishes a scale-free relation between structure constant

$$C_n^2 = \frac{2\Gamma(2 - \nu)C_s}{(4\pi)^{3/2}\,(\nu - 1)\,\Gamma(\nu + 1/2)2^{2\nu-2}}\qquad(3.29)$$

and the turbulent strength C_s.

In turbulence theory it is customary to place the $(2\pi)^3$ factor in the spatial-domain transform rather than the spectral domain. To avoid confusion let $\overline{C}_s^2 = C_s/(2\pi)^3$ where the overbar denotes the constant as it is usually reported in turbulence theory. For isotropic Kolmogorov turbulence $\Phi_n(q) \sim C_s q^{-11/3}$, which corresponds to $\nu = 4/3$. The relation

$$C_s = \frac{(4\pi)^{3/2}\,(\nu - 1)\,\Gamma(\nu + 1/2)2^{2\nu-2}}{2\Gamma(2 - \nu)}\overline{C}_n^2$$

$$= 0.033\overline{C}_n^2\qquad(3.30)$$

is commonly used for atmospheric turbulence [5, Appendix B4].

In summary, the spectral characterization of isotropic turbulence admits a power-law form characterized by a scale-free turbulent strength parameter and a spectral index. The structure function admits a complementary power-law form scaled by the structure constant.   Figures 3.2 and 3.3 show the SDF and structure function for Kolmogorov turbulence, respectively.   The complementary power-law approximations are superimposed as dashed curves.

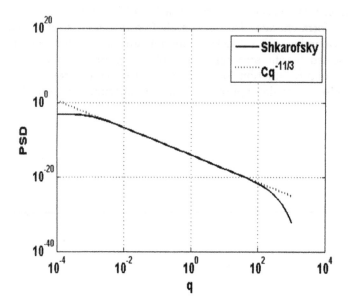

Figure 3.2 Shkarofsky form of radial wavenumber SDF with outer and inner scale cutoffs (solid), and Kolmogorov power-law approximation (dashed).

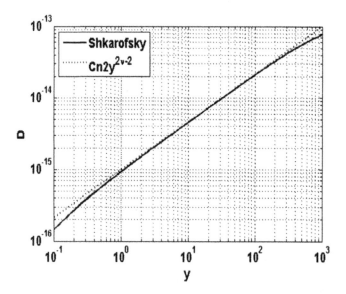

Figure 3.3 Shkarofsky form of radial wavenumber structure function with outer and inner scale cutoffs (solid), and Kolmogorov power-law approximation (dashed).

3.1.5   Phase Structure

Although the FPE formally responds to the in situ structure directly, the phenomenon captured by the split-step solution exploits the more gradual development of intensity scintillation. In effect, the initiating structure is mapped onto the wavefront phase as an integral over a segment of length l_p.[10] The local media interaction maps onto the evolving field as a multiplicative correction of the form

$$u(\varsigma) = \exp\left\{ik \int_0^{l_p} \delta n(\eta, \varsigma) d\eta\right\}.$$ (3.31)

This integration collapses the relevant phase structure from three to two dimensions. Assuming that δn is Gaussian, it follows that

$$\langle u(\varsigma) u^*(\varsigma')\rangle = \exp\left\{-k^2 \frac{1}{2} \left\langle \left[\int_0^{l_p} (\delta n(\eta, \varsigma) - \delta n(\eta, \varsigma')) \, d\eta\right]^2\right\rangle\right\}$$

$$= \exp\left\{-k^2 \int_0^{l_p} \int_0^{l_p} [R_n(\eta - \eta', 0) - R_n(\eta - \eta', \Delta\varsigma)] \, d\eta d\eta'\right\}$$

$$= \exp\left\{-k^2 l_p \int_{-l_p}^{l_p} (1 - |\Delta\eta|/l_p) \, (D_n(\Delta\eta, \Delta\varsigma)\right.$$

$$\left. - D_n(\Delta\eta, 0)) \, d\Delta\eta\right\},$$ (3.32)

and

$$R_u(\Delta\varsigma) = \exp\left\{-k^2 l_p \int_{-\infty}^{\infty} (D_{\delta n}(\Delta\eta, \Delta\varsigma) - D_{\delta n}(\Delta\eta, 0)) \, d\Delta\eta\right\}.$$ (3.33)

To evaluate this integral, the integrand of integral in the argument of (3.33) is written explicitly. As a building block

$$R_{\delta\phi}(\Delta\varsigma) = k^2 l_p \int_{-\infty}^{\infty} R_{\delta n}(\Delta\eta, \Delta\varsigma) d\Delta\eta$$

$$= k^2 l_p \iint \Phi_{\delta n}(\kappa, 0) \exp\{i\kappa \cdot \Delta\varsigma\} d\kappa/(2\pi)$$

$$= k^2 l_p \int_0^{\infty} J_o(\kappa\Delta\varsigma)\Phi_{\delta n}(\kappa, 0)\kappa d\kappa/(2\pi).$$ (3.34)

---

[10]See (2.22).

Using the Shkarofsky spectral density function, it follows that

$$
\begin{aligned}
\Re_{\delta\phi}(y) &= R_{\delta\phi}(y)/R_{\delta\phi}(0) \\
&= \int_0^\infty J_o(\kappa y)Q(\kappa)\kappa d\kappa/(2\pi)/\varkappa \\
&= \frac{\left(\sqrt{1+(yq_S)^2}\right)^{\nu-1/2} K_{\nu-1/2}\left(2\frac{q_L}{q_S}\sqrt{1+(yq_S)^2}\right)}{K_{\nu-1/2}\left(2\frac{q_L}{q_S}\right)}.
\end{aligned}
\tag{3.35}
$$

The normalization $\varkappa$ is introduced to makes $\Re_{\delta\phi}(0) = 1$. It can be rewritten and evaluated as

$$
\begin{aligned}
\varkappa &= \int_0^\infty Q(q)qdq/(2\pi) \\
&= \frac{(4\pi/(q_S q_L))^{1/2} K_{\nu-1/2}(2q_L/q_S)}{K_{\nu-1}(2q_L/q_S)} \\
&\sim \frac{\Gamma(\nu-1/2)}{(4\pi)^{1/2}q_L\Gamma(\nu-1)}.
\end{aligned}
\tag{3.36}
$$

The path-integrated phase variance is

$$
R_{\delta\phi}(0) = k^2 l_p \left\langle \delta n^2 \right\rangle \varkappa.
\tag{3.37}
$$

The two-dimensional phase spectrum is related to the three-dimensional refractive index spectrum by scaling and projection

$$
\begin{aligned}
\Phi_{\delta\phi}(q) &= k^2 l_p \left\langle \delta n^2 \right\rangle Q(q) \\
&\sim k^2 l_p C_s \left(q_L^2 + q^2\right)^{-(\nu+1/2)}.
\end{aligned}
\tag{3.38}
$$

The phase variance $\left\langle \delta\phi^2 \right\rangle = R_{\delta\phi}(0)$ is

$$
\begin{aligned}
\left\langle \delta\phi^2 \right\rangle &= k^2 l_p \left\langle \delta n^2 \right\rangle \kappa \\
&\sim k^2 l_p \frac{\Gamma(\nu-1)C_s}{(4\pi)^{3/2}\Gamma(\nu+1/2)q_L^{2\nu-2}}(4\pi)^{1/2}\frac{\Gamma(\nu-1/2)}{q_L\Gamma(\nu-1)} \\
&\sim k^2 l_p q_L^{-1}\frac{\Gamma(\nu-1/2)C_s}{4\pi\Gamma(\nu+1/2)q_L^{2\nu-2}}.
\end{aligned}
\tag{3.39}
$$

The phase structure function is manipulated as with its three-dimensional form:

$$
\begin{aligned}
D_{\delta\phi}(y) &= \frac{k^2 l_p C_s \Gamma(\nu-1/2)}{2\pi\Gamma(\nu+1/2)} \\
&\quad \times \frac{1 - 2|q_L y/2|^{\nu-1/2} K_{\nu-1/2}(q_L y)/\Gamma(\nu-1/2)}{q_L^{2\nu-1}} \\
&\sim \frac{k^2 l_p C_s \Gamma(3/2-\nu)}{\pi\Gamma(\nu+1/2)(2\nu-1)2^{2\nu-1}}|y|^{2\nu-1} \\
&= C_{\delta\phi}|y|^{2\nu-1},
\end{aligned}
\tag{3.40}
$$

where
$$C_{\delta\phi} = \frac{k^2 l_p C_s \Gamma(3/2 - \nu)}{\pi \Gamma(\nu + 1/2)(2\nu - 1)2^{2\nu-1}}. \tag{3.41}$$

Note that the valid range for the phase structure constant is $0.5 < \nu < 1.5$, which has been shifted from $1 < \nu < 2$ for the in situ turbulence structure constant in (3.29). If $u(\zeta) = \exp\{-i\delta\phi(\zeta)\}$, and $\delta\phi(\zeta)$ is a homogeneous Gaussian process, the mutual coherence function satisfied the following relation:

$$
\begin{aligned}
R_u(\Delta\zeta) &= \exp\{-D_{\delta\phi}(\Delta\zeta)\} \\
&\simeq \exp\left\{-C_{\delta\phi}|\Delta\zeta|^{2\nu-1}\right\}. \tag{3.42}
\end{aligned}
$$

3.1.6 Anisotropy

Up to this point only isotropic irregularity structures have been considered. However, the conditions under which structure develops can generate a persistent preferential direction in the evolving structure. Moreover, in magnetically biased plasmas charged particles move much more freely along background magnetic field lines than across magnetic field lines. This leads to field-aligned anisotropic irregularity structures. Singleton [43] introduced an efficient method to incorporate preferential correlation directions. In isotropic models, the correlation and spectral functions depend only on the magnitude of the difference vector or the magnitude of the wavenumber. Surfaces of constant correlation or spectral intensity are spherical. Singleton's scheme uses coordinate transformations that can deform and translate the contours into isotropic form. Because the manipulations are purely algebraic there is no loss of generality in proceeding with the unmodified isotropic model. The appropriate transformations will be introduced in Chapter 4 where the effects of anisotropy are essential.

3.2 CALCULATION OF FIELD MOMENTS

A formal statistical characterization of the complex field, $\psi(x, \zeta_n)$, follows from the moments

$$\Gamma_{nm}(x; \zeta_1, \cdots \zeta_N; \xi_1, \cdots \xi_M) = \left\langle \prod_{n=1}^{N} \psi(x, \zeta_n) \prod_{m=1}^{M} \psi^*(x, \xi_m) \right\rangle, \tag{3.43}$$

which are functions of x and the $N + M$ transverse spatial variables. The following notation will be used in the development:

$$P_{NM}(x; \zeta_1, \cdots \zeta_N; \xi_1, \cdots \xi_M) = \prod_{n=1}^{N} \psi(x, \zeta_n) \prod_{m=1}^{M} \psi^*(x, \xi_m). \tag{3.44}$$

The strategy is to use the FPE approximation

$$\psi(x + \Delta x, \zeta) = \psi(x, \zeta) + ik\Theta\psi(x, \zeta)\Delta x + ik\delta n(x, \zeta)\psi(x, \zeta)\Delta x, \qquad (3.45)$$

to compute $\Gamma_{NM}(x + \Delta x; \zeta_1, \cdots \zeta_N; \xi_1, \cdots \xi_M)$ first by retaining only terms that are linear in Δx. To the extent that the expanded terms can be manipulated to isolate the field moments as products, a first-order differential equation results. Substituting (3.45) into (3.43) and retaining only terms that are linear in Δx, the following intermediate result for the moments with $M = N$ can be obtained:

$$\frac{\partial \Gamma_{NN}(\cdots)}{\partial x} =$$

$$-ik_1 \sum_{k=1}^{N} \langle \Theta_{\zeta_k} P_{NN}(\cdots) \rangle + ik_2 \sum_{l=1}^{N} \langle \Theta_{\xi_l} P_{NN}(\cdots) \rangle$$

$$- \left\langle \sum_{n=1}^{N} (ik_1 \delta n(x, \zeta_n) - ik_2 \delta n(x, \xi_n)) P_{NN}(\cdots) \right\rangle. \qquad (3.46)$$

The ellipsis is replaced by the argument in (3.43). The propagation operator notation Θ_{ζ_n} means the propagation operator is applied to the variable ζ_n. The diffraction term can be put in the desired product form by using a standard property of the expectation of derivatives of random variables. Because Θ_{ζ_n} is formally defined by a Taylor series of derivative operations, it follows that

$$\langle \Theta_{\zeta_k} P_{NN}(\cdots) \rangle = \Theta_{\zeta_k} \Gamma_{NN}(\cdots). \qquad (3.47)$$

The term involving products of the refractive-index and products of fields that depend on the refractive index is difficult to evaluate. The Navikov-Furutsu Theorem [5, Chapter 20.4] provides the following relationship for a real Gaussian process $\delta n(\mathbf{r})$

$$\langle \delta n(\mathbf{r}) Z(\mathbf{r}, \mathbf{r}') \rangle = \iiint \langle \delta n(\mathbf{r}) \delta n(\mathbf{r}'') \rangle \left\langle \frac{\partial Z(\mathbf{r}, \mathbf{r}')}{\partial \delta n(\mathbf{r}'')} \right\rangle dV'', \qquad (3.48)$$

but it doesn't provide a tractable answer directly because the expectation of a functional (Gateaux) derivative [44, Chapter 3.4], $\partial Z(\mathbf{r}, \mathbf{r}')/\partial \delta n(\mathbf{r}'')$, must be evaluated. The result, which is singular, reduces the three-dimensional integral to a product involving path-integrated quantities. The condition that affects this translation is referred to as the Markov approximation. The terminology may be confusing because the forward approximation is intrinsically Markovian in that knowledge of the field at x under the forward approximation is sufficient to predict the field in any plane beyond x. As it is used to evaluate the expectation of the functional derivative, the Markov approximation requires structure along the path to decorrelate before any significant field change occurs.

In any case, a translation of the correlated product terms in (3.46) to the desired product form is achieved by using the formal relation

$$\langle \delta n(\mathbf{r}) P_{NN}(\cdots) \rangle = \langle \delta \overline{n}(\zeta) \delta \overline{n}(\zeta') \rangle \Gamma_{NN}(\cdots), \tag{3.49}$$

where the notation $\delta \overline{n}(\zeta)$ means per unit of path integrated refractive index perturbation. Upon applying (3.49) and (3.46), the following first-order differential equation for the field moments is obtained:

$$\frac{\partial \Gamma_{NN}(\cdots)}{\partial x} = -ik_1 \sum_{k=1}^{N} \Theta_{\zeta_k} \Gamma_{NN}(\cdots) + ik_2 \sum_{l=1}^{N} \Theta_{\xi_l} \Gamma_{NN}(\cdots)$$

$$- \left\langle \left(\sum_{n=1}^{N} (ik_1 \delta \overline{n}(\zeta_n) - ik_2 \delta \overline{n}(\xi_n)) \right)^2 \right\rangle \Gamma_{NN}(\cdots). \tag{3.50}$$

Expanding and manipulating the media-interaction term shows that

$$- \left\langle \left(\sum_{n=1}^{N} (ik_1 \delta \overline{n}(\zeta_n) - ik_2 \delta \overline{n}(\xi_n)) \right)^2 \right\rangle =$$

$$+ \frac{1}{2} N^2 R_{\delta \overline{n}}(0)(k_1 - k_2)^2$$

$$- \frac{1}{2} \sum_{n=1}^{N} \sum_{m=1}^{N} D_{\delta \overline{n}}(\zeta'_n - \zeta_m) k_1^2$$

$$- \frac{1}{2} \sum_{n=1}^{N} \sum_{m=1}^{N} D_{\delta \overline{n}}(\xi'_n - \xi_m) k_2^2$$

$$+ \frac{1}{2} \sum_{n=1}^{N} \sum_{m=1}^{N} D_{\delta \overline{n}}(\zeta'_n - \xi_m) k_1 k_2$$

$$+ \frac{1}{2} \sum_{n=1}^{N} \sum_{m=1}^{N} D_{\delta \overline{n}}(\xi'_n - \zeta_m) k_1 k_2. \tag{3.51}$$

A complete derivation of (3.50) has been presented by L. C. Lee, who used characteristic functionals in his development of moment equations [34]. Lee's result is presented in terms of autocorrelation functions rather than structure functions, but it is otherwise equivalent to the form presented here. Note that the product form that allows the split-step solution to the FPE is preserved by (3.49).

3.3 SECOND-ORDER MOMENTS

The two-frequency second-order moment equation can be written as

$$
\begin{aligned}
\frac{\partial \Gamma_{11}(x,\boldsymbol{\xi},\boldsymbol{\xi}';k_1,k_2)}{\partial x} &= -i\left(k_1\Theta_{\boldsymbol{\xi}} - k_2\Theta_{\boldsymbol{\xi}}'\right)\Gamma_{11}(x,\boldsymbol{\xi},\boldsymbol{\xi}';k_1,k_2) \\
&\quad - \left(R_{\delta\overline{n}}(0)\frac{(k_1-k_2)^2}{2} + D_{\delta\overline{n}}(\boldsymbol{\xi}-\boldsymbol{\xi}')k_1k_2\right) \\
&\quad \times \Gamma_{11}(x,\boldsymbol{\xi},\boldsymbol{\xi}';k_1,k_2).
\end{aligned}
\tag{3.52}
$$

The split-step strategy can be used again to formulate solutions to (3.52). In the absence of diffraction, the media-interaction term can be integrated directly. The single phase screen model requires only the evaluation of a freely propagating wave field. If $k_1 = k_2$, the irregularity structure is homogeneous, and the excitation field is a plane wave, hence the diffraction terms cancel. In this case, the complete solution has no variation beyond the region of interaction with the medium:

$$
\Gamma_{11}(x,\boldsymbol{\Delta\xi};k) = \exp\left\{-l_p k^2 D_{\delta\overline{n}}(\boldsymbol{\Delta\xi})\right\}\Gamma_{11}(0,\boldsymbol{\Delta\xi};k).
\tag{3.53}
$$

The same result can be obtained by assuming the diffraction process was initiated with a Gaussian phase screen with structure function $D_{\delta\overline{n}}(\boldsymbol{\Delta\xi})$.

Now consider the more general two-frequency component of (3.52) for propagation in free space,

$$
\frac{\partial \Gamma_{11}(x,\boldsymbol{\xi},\boldsymbol{\xi}';k_1,k_2)}{\partial x} = -i\left(k_1\Theta_{\boldsymbol{\zeta}} - k_2\Theta_{\boldsymbol{\xi}}\right)\Gamma_{11}(x,\boldsymbol{\xi},\boldsymbol{\xi}';k_1,k_2).
\tag{3.54}
$$

First the generalized Fourier transform

$$
\begin{aligned}
\widehat{\Gamma}_{11}(x,\boldsymbol{\kappa},\boldsymbol{\kappa}';k_1,k_2) &= \iint\iint \Gamma_{11}(x,\boldsymbol{\xi},\boldsymbol{\xi}';k_1,k_2) \\
&\quad \times \exp\left\{-i\left(\boldsymbol{\kappa}\cdot\boldsymbol{\xi} + \boldsymbol{\kappa}'\cdot\boldsymbol{\xi}'\right)\right\}d\boldsymbol{\xi}d\boldsymbol{\xi}',
\end{aligned}
\tag{3.55}
$$

is introduced. It follows by direct computation that

$$
\frac{\partial \widehat{\Gamma}_{11}(x,\boldsymbol{\kappa},\boldsymbol{\kappa}';k_1,k_2)}{\partial x} = -i\left(k_1 g(\boldsymbol{\kappa}) - k_2 g(\boldsymbol{\kappa}')\right)\widehat{\Gamma}_{11}(\boldsymbol{\kappa},\boldsymbol{\kappa}';k_1,k_2).
\tag{3.56}
$$

This result now can be integrated directly to obtain the form

$$
\widehat{\Gamma}_{11}(x,\boldsymbol{\kappa},\boldsymbol{\kappa}';k_1,k_2) = \exp\left\{-i\left(k_1 g_1(\boldsymbol{\kappa}) - k_2 g_2(\boldsymbol{\kappa}')\right)x\right\}\widehat{\Gamma}_{11}(\boldsymbol{\kappa},\boldsymbol{\kappa}';k_1,k_2).
\tag{3.57}
$$

Following the split-step procedure, let

$$\Gamma_{11}(\Delta\xi; k_1, k_2) = \exp\left\{-l_p R_{\delta\bar{n}}(0)(k_1 - k_2)^2/2 - l_p D_{\delta\bar{n}}(\Delta\xi)k_1 k_2\right\}. \quad (3.58)$$

Because the spatial variation depends only on $\Delta\xi = \xi - \xi'$, the generalized Fourier transform can be evaluated explicitly to obtain the result:

$$
\begin{aligned}
\widehat{\Gamma}_{11}(\kappa, \kappa'; k_1, k_2) &= \exp\left\{-l_p R_{\delta\bar{n}}(0)(k_1 - k_2)^2/2\right\} \\
&\times (2\pi)^2 \delta(\kappa - \kappa') \iint \exp\left\{-l_p D_{\delta\bar{n}}(\Delta\xi)k_1 k_2\right\} \\
&\times \exp\left\{-i\kappa\cdot\Delta\xi\right\} d\Delta\xi.
\end{aligned}
\quad (3.59)
$$

Substituting (3.59) into (3.52), it follows that

$$
\begin{aligned}
\widehat{\Gamma}_{11}(x, \kappa, \kappa'; k_1, k_2) &= \exp\left\{-l_p R_{\delta\bar{n}}(0)(k_1 - k_2)^2/2\right\}(2\pi)^2 \delta(\kappa - \kappa') \\
&\times \iint \exp\left\{-l_p D_{\delta\bar{n}}(\Delta\xi)k_1 k_2\right\} \\
&\times \exp\left\{-i\left(k_1 g_1(\kappa) - k_2 g_2(\kappa')\right)x\right\} \\
&\times \exp\left\{-i\kappa\cdot\Delta\xi\right\} d\Delta\xi.
\end{aligned}
\quad (3.60)
$$

A transformation back to the spatial domain gives the final form for the two-frequency mutual-coherence function:

$$
\begin{aligned}
\Gamma_{11}(x; \Delta\xi; k_1, k_2) &= \exp\left\{-l_p R_{\delta\bar{n}}(0)(k_1 - k_2)^2/2\right\} \\
&\times \iint \exp\left\{-l_p D_{\delta\bar{n}}(\Delta\xi)k_1 k_2\right\} \\
&\times \iint \exp\left\{-i\left(k_1 g_1(\kappa) - k_2 g_2(\kappa)\right)x\right\} \\
&\times \exp\left\{-i\kappa\cdot\left(\Delta\xi' - \Delta\xi\right)\right\}\frac{d\kappa}{(2\pi)^2}d\Delta\xi' \quad (3.61)
\end{aligned}
$$

Note that up to this point the narrow-angle scatter approximation has not been used. However, considerable simplification is realized when the narrow-angle scatter approximation $g_j(\kappa) \simeq 1 - \kappa^2/(2k_j)$ is used. Following the development of the parabolic wave equation in Chapter 2, the leading 1 in the approximation can be removed by introducing multiplicative terms of the form $\exp\{ik_j x\}$, which can be absorbed in a redefinition of the covariance function. Once this has been done, the integral over κ can be evaluated. The more analytically tractable form involving only a single integration is

$$
\begin{aligned}
\Gamma_{11}(x; \Delta\xi; k_1, k_2) &= \exp\left\{-l_p R_{\delta\bar{n}}(0)(k_1 - k_2)^2/2\right\} \\
&\times \iint \exp\left\{-l_p D_{\delta\bar{n}}(\Delta\xi)k_1 k_2\right\} \\
&\times \frac{\exp\left\{-\left(\Delta\xi' - \Delta\xi\right)^2/\alpha\right\}}{\pi\alpha}d\Delta\xi', \quad (3.62)
\end{aligned}
$$

where

$$\alpha = 2i\left(1/k_1 - 1/k_2\right)x. \tag{3.63}$$

The single-point correlation obtained with $\Delta\boldsymbol{\xi} = 0$ is the same as Equation (12) in Rino et al. [45] and references cited therein. Discussion of the ramifications of loss of frequency coherence will be deferred to Chapter 5. The remaining analysis in this chapter will consider only single-frequency measurements.

3.4   FOURTH-ORDER MOMENTS

The intensity correlation function, $\langle I(x,\boldsymbol{\zeta})I(x,\boldsymbol{\xi})\rangle$, the scintillation index, SI, which was introduced in Chapter 1, and the intensity SDF all follow from the single-frequency fourth-order moment $\Gamma_{22}(x,\boldsymbol{\zeta}_1,\boldsymbol{\zeta}_2;\boldsymbol{\xi}_1,\boldsymbol{\xi}_2)$. For intensity correlations the parabolic approximation is incorporated by the replacement $\Theta_{\boldsymbol{\zeta}} \simeq 1 + \nabla^2_{\boldsymbol{\zeta}_1}/\left(2k^2\right)$. The phase corrections that remove the leading 1 cancel, whereby only the transverse Laplacian enters the simplified propagation operator. The following form of the fourth-order moment equation follows from (3.50) and (3.51):

$$\frac{\partial\Gamma_{22}(x,\boldsymbol{\zeta}_1,\boldsymbol{\zeta}_2;\boldsymbol{\xi}_1,\boldsymbol{\xi}_2)}{\partial x} = -\frac{i}{2k}\left[\nabla^2_{\boldsymbol{\zeta}_1} + \nabla^2_{\boldsymbol{\zeta}_2} - \nabla^2_{\boldsymbol{\xi}_1} - \nabla^2_{\boldsymbol{\xi}_1}\right]\Gamma_{2,2}(x,\cdots)$$
$$-\frac{k}{2}\left[D_{\delta\overline{\phi}}(\boldsymbol{\zeta}_1 - \boldsymbol{\xi}_2) + D_{\delta\overline{\phi}}(\boldsymbol{\zeta}_2 - \boldsymbol{\xi}_1) + D_{\delta\overline{\phi}}(\boldsymbol{\zeta}_2 - \boldsymbol{\xi}_2) + \right.$$
$$\left. D_{\delta\overline{\phi}}(\boldsymbol{\zeta}_1 - \boldsymbol{\xi}_1) - D_{\delta\overline{\phi}}(\boldsymbol{\zeta}_1 - \boldsymbol{\zeta}_2) - D_{\delta\overline{\phi}}(\boldsymbol{\xi}_1 - \boldsymbol{\xi}_2)\right]\Gamma_{2,2}(x,\cdots). \tag{3.64}$$

For single-frequency measurements, the equivalence $\delta\overline{\phi} = k\delta\overline{n}$ simplifies the notation. Also, although it is not strictly necessary here, the 1 in the diffraction operator can be dropped with the understanding that the reference $\exp\{ikx\}$ has been removed. With the variable transformation

$$\boldsymbol{\alpha}_0 = \frac{1}{4}(\boldsymbol{\zeta}_1 + \boldsymbol{\zeta}_2 + \boldsymbol{\xi}_1 + \boldsymbol{\xi}_2) \tag{3.65}$$

$$\boldsymbol{\alpha} = (\boldsymbol{\zeta}_1 + \boldsymbol{\zeta}_2 - \boldsymbol{\xi}_1 - \boldsymbol{\xi}_2) \tag{3.66}$$

$$\boldsymbol{\alpha}_1 = \frac{1}{2}(\boldsymbol{\zeta}_1 - \boldsymbol{\zeta}_2 + \boldsymbol{\xi}_1 - \boldsymbol{\xi}_2) \tag{3.67}$$

$$\boldsymbol{\alpha}_2 = \frac{1}{2}(\boldsymbol{\zeta}_1 - \boldsymbol{\zeta}_2 - \boldsymbol{\xi}_1 + \boldsymbol{\xi}_2) \tag{3.68}$$

(3.64) can be simplified further to

$$\frac{\partial\Gamma_{22}(x,\boldsymbol{\alpha}_0,\boldsymbol{\alpha};\boldsymbol{\alpha}_1,\boldsymbol{\alpha}_2)}{\partial x} = \frac{i}{k}\left[\nabla_{\boldsymbol{\alpha}_0}\cdot\nabla_{\boldsymbol{\alpha}} - \nabla_{\boldsymbol{\alpha}_1}\cdot\nabla_{\boldsymbol{\alpha}_2}\right]\Gamma_{22}(x,\cdots) + $$
$$k\left[D_{\delta n}(\boldsymbol{\alpha}_1 + \boldsymbol{\alpha}/2) + D_{\delta n}(\boldsymbol{\alpha}_1 - \boldsymbol{\alpha}/2) + D_{\delta n}(\boldsymbol{\alpha}_2 - \boldsymbol{\alpha}/2) + \right.$$
$$\left. D_{\delta n}(\boldsymbol{\alpha}_2 + \boldsymbol{\alpha}/2) - D_{\delta n}(\boldsymbol{\alpha}_1 + \boldsymbol{\alpha}_2) - D_{\delta n}(\boldsymbol{\alpha}_1 - \boldsymbol{\alpha}_2)\right]\Gamma_{22}(x,\cdots). \tag{3.69}$$

Because the media-interaction term does not depend on α_0,

$$\Gamma_{22}(x, \alpha_0, \alpha; \alpha_1, \alpha_2)$$

will not depend on α_0 unless the initial intensity covariance depends on α_0. Furthermore, an adjustment to the elements of α_0 will render $\alpha = 0$. Without loss of generality, the final form of the fourth-order moment equation becomes

$$\frac{\partial \Gamma_{22}(x, \alpha_1, \alpha_2)}{\partial x} = -\frac{i}{k} \nabla_{\alpha_2} \cdot \nabla_{\alpha_3} \Gamma_{22}(x, \alpha_1, \alpha_2) -$$
$$\frac{k}{2} \left[2D_{\delta\bar{\phi}}(\alpha_1) + 2D_{\delta\bar{\phi}}(\alpha_2) - \right.$$
$$\left. D_{\delta\bar{\phi}}(\alpha_2 + \alpha_2) - D_{\delta\bar{\phi}}(\alpha_1 - \alpha_2) \right] \Gamma_{22}(x, \alpha_1, \alpha_2). \qquad (3.70)$$

First consider the solution to (3.70) in the absence of diffraction:

$$\Gamma_{22}(x, \alpha_1, \alpha_2) = \exp \left\{ -ikx \left[D_{\delta\bar{\phi}}(\alpha_1) + D_{\delta\bar{\phi}}(\alpha_2) \right. \right.$$
$$\left. \left. -D_{\delta\bar{\phi}}(\alpha_1 + \alpha_2)/2 - D_{\delta\bar{\phi}}(\alpha_1 - \alpha_2)/2 \right] \right\}. \quad (3.71)$$

As with the second-order moments this is the same result that would be obtained if a Gaussian phase screen had been invoked. The symmetric disposition of the phase structure function components is exploited to obtain

$$D_{\delta\bar{\phi}}(\alpha_1) + D_{\delta\bar{\phi}}(\alpha_2) - D_{\delta\bar{\phi}}(\alpha_1 + \alpha_2)/2 - D_{\delta\bar{\phi}}(\alpha_1 - \alpha_2)/2$$
$$= 8 \iint \Phi_{\delta n}(\boldsymbol{\kappa}, 0) \sin^2 (\boldsymbol{\kappa} \cdot \alpha_1/2) \sin^2 (\boldsymbol{\kappa} \cdot \alpha_2/2) \frac{d\boldsymbol{\kappa}}{(2\pi)^2}. \qquad (3.72)$$

The singular behavior of the power-law spectrum $\kappa \Phi_{\delta n}(\boldsymbol{\kappa}, 0) \sim \kappa^{-2v}$ is compensated by the product of the sin^2 terms, which approach zero as κ^8.

Pursuing the split-step strategy again, consider the solution to (3.70) in the absence of phase structure:

$$\frac{\partial \Gamma_{22}(x, \alpha_1, \alpha_2)}{\partial x} = -\frac{i}{k} \nabla_{\alpha_1} \cdot \nabla_{\alpha_2} \Gamma_{22}(x, \alpha_1, \alpha_2). \qquad (3.73)$$

Introducing the generalized Fourier transformation

$$\widehat{\Gamma}_{22}(x, \boldsymbol{\kappa}_1, \boldsymbol{\kappa}_2) = \iint \iint \Gamma_{22}(x, \alpha_1, \alpha_2) \exp \left\{ -i (\boldsymbol{\kappa}_1 \cdot \alpha_1 + \boldsymbol{\kappa}_2 \cdot \alpha_2) \right\} d\alpha_1 d\alpha_2, \qquad (3.74)$$

It follows from (3.73) that

$$\frac{\partial \widehat{\Gamma}_{22}(x, \boldsymbol{\kappa}_1, \boldsymbol{\kappa}_2)}{\partial x} = \frac{i}{k} \boldsymbol{\kappa}_1 \cdot \boldsymbol{\kappa}_2 \widehat{\Gamma}_{22}(x, \boldsymbol{\kappa}_1, \boldsymbol{\kappa}_2). \qquad (3.75)$$

If $\widehat{\Gamma}_{22}(0, \kappa_1, \kappa_2)$ is known, the generalized Fourier spectrum in the plane located at x is

$$\widehat{\Gamma}_{22}(x, \kappa_1, \kappa_2) = \exp\{i\kappa_1 \cdot \kappa_2 x/k\}\widehat{\Gamma}_{22}(0, \kappa_1, \kappa_2). \tag{3.76}$$

These results can now be used to compute the intensity correlation function for a single phase screen, namely

$$\begin{aligned}
\langle I(\boldsymbol{\xi}_1)I(\boldsymbol{\xi}_2)\rangle &= \Gamma_{22}(x, \boldsymbol{\xi}_1, \boldsymbol{\xi}_2; \boldsymbol{\xi}_1, \boldsymbol{\xi}_2) \\
&= \Gamma_{22}(x, \boldsymbol{\alpha}_1; \mathbf{0}). \tag{3.77}
\end{aligned}$$

The corresponding intensity SDF is given by the integral

$$\Phi_I(\kappa_1) = \iint \Gamma_{22}(x, \boldsymbol{\alpha}_1; \mathbf{0}) \exp\{-i\kappa_1 \cdot \boldsymbol{\alpha}_1\}\, d\boldsymbol{\alpha}_1. \tag{3.78}$$

For a single phase screen with thickness l_p, it follows from (3.32) that

$$\begin{aligned}
\Gamma_{22}(0, \boldsymbol{\alpha}_1, \boldsymbol{\alpha}_2) &= \exp\big\{-ik^2 l_p\, [D_{\delta\phi}(\boldsymbol{\alpha}_1) + D_{\delta\phi}(\boldsymbol{\alpha}_2) \\
&\quad - D_{\delta\phi}(\boldsymbol{\alpha}_1 + \boldsymbol{\alpha}_2)/2 - D_{\delta\phi}(\boldsymbol{\alpha}_1 - \boldsymbol{\alpha}_2)/2]\big\}. \tag{3.79}
\end{aligned}$$

The SDF in the plane at distance x is

$$\begin{aligned}
\widehat{\Gamma}_{22}(x, \kappa_1, \kappa_2) &= \iint \iint \Gamma_{22}(0, \boldsymbol{\alpha}_1', \boldsymbol{\alpha}_2') \\
&\quad \times \exp\{-i\,(\kappa_1 \cdot \boldsymbol{\alpha}_1' + \kappa_2 \cdot \boldsymbol{\alpha}_2')\}d\boldsymbol{\alpha}_1' d\boldsymbol{\alpha}_2' \\
&\quad \times \exp\{i\kappa_1 \cdot \kappa_2 x/k\}. \tag{3.80}
\end{aligned}$$

The intensity SDF now can be calculated directly from the fourth-order moment as

$$\begin{aligned}
\Phi_I(\kappa_1) &= \iint \Gamma_{22}(x, \boldsymbol{\alpha}_1'; \mathbf{0}) \exp\{-i\kappa_1 \cdot \boldsymbol{\alpha}_1'\}\, d\boldsymbol{\alpha}_1' \\
&= \iint \widehat{\Gamma}_{22}(x, \kappa_1, \kappa_2') \frac{d\kappa_2'}{(2\pi)^2} \\
&= \iint \Gamma_{22}(0, \boldsymbol{\alpha}_1, \kappa_1 x/k) \exp\{-i\kappa_1 \cdot \boldsymbol{\alpha}_1\} d\boldsymbol{\alpha}_1. \tag{3.81}
\end{aligned}$$

3.4.1 Solutions to the Fourth-Order Moment Equation

The equations (3.70) and (3.81), which characterize the fourth-order moments of a field propagating in a power-law environment, have been known since the early 1970s. The first numerical solution to (3.70) was published by Yeh, Liu, and Youakim [46]. The citation is to a later publication in the January, 1975 issue of *Radio Science*, which was devoted to propagation in continuous random media. The issue also contains a published numerical solution to (3.81) by Marians [47], which is based on an analytic solution to the power-law form (3.72) developed by Rumsey [48]. The Rumsey-Marians result exploits a comparatively simple form for the intensity spectrum that had been known and used for some time ([49], [50], [51], [52], [53]).

The early development of scintillation theory was influenced by diagrams based on calculations that used a Gaussian functional form for the irregularity autocorrelation function. The Gaussian form has a single dominant scale size rather than a power-law spatial-wavenumber distribution. Based on results developed initially by Salpeter [54] and Singleton [43] organized scintillation in a diagram in which the Fresnel scale $\sqrt{x/k}$ as normalized to the correlation scale formed the vertical dimension and the RMS phase imposed by transiting the disturbed region formed the horizontal dimension. In the Fraunhofer region, the Fresnel scale is smaller than the correlation scale. When the RMS phase is large enough to produce strong scintillation in the Fraunhofer region, the scintillation index exceeds unity. In the Fresnel region, the Fresnel scale exceeds the correlation scale. The increase in the scintillation index from strong to weak scatter in the Fresnel region approaches unity monotonically. This result is consistent with Mercier's [20] result, which shows that the in-phase and quadrature-phase variance become equal at arbitrarily large distances from a phase screen.

Many of the papers in the January 1975 issue of *Radio Science* and publications that preceded them discuss strong scatter and the conditions under which strong focusing $(SI > 1)$ occurs. Intensity correlation and SDF characteristics were also discussed. This early work stimulated a large number of researchers to pursue the characterization of scintillation under strong-scatter conditions. This activity culminated with a number of exceptional publications a decade or so later. The review paper by Codona, et. al. [55] is particularly informative. The results are based on moment expansions and path-integral methods that exploit the similarity between the parabolic wave equation and equations of theoretical physics that characterize probabilistic wave functions and potential fields [56]. Uscinski [57] pursued a similar methodology, which explicitly develops a multiple phase screen method that resembles the split-step method utilized in the development above. Another class of solutions is based on the two-scale method developed by Whitman and Beran [58], [59], [60]. The two-scale method has produced approximate analytic forms that capture the essential elements of the scintillation index behavior and the structure of the spectral density function. Other references from this period include two-frequency intensity correlations [61], [62], [63], [64]. More recently numerical simulations have emerged as viable research tools that have further extended the range of problems that have been studied [65], [66]. The eikonal method, which replaces the complex field by a form that includes amplitude and phase explicitly is also used routinely for analytical moment computation [67], [68], [69].

3.4.2   Power-Law Scintillation Regimes

To summarize the general scattering characteristics in power-law environments, the phase screen model will be used. Following the development in Rino 1979 [70], (3.81) is rewritten as

$$\Phi_I(\kappa) = \iint \exp\left\{-g(\alpha, \kappa x/k)\right\} \exp\{-i\kappa \cdot \alpha\} d\alpha, \qquad (3.82)$$

where

$$g(\xi, \eta) = k^2 l_p \left[D_{\delta\bar{n}}(\xi) + D_{\delta\bar{n}}(\eta) - D_{\delta\bar{n}}(\xi + \eta)/2 - D_{\delta\bar{n}}(\xi - \eta)/2\right]. \qquad (3.83)$$

Inserting the power-law form of the phase spectrum into (3.72) and changing variables shows that

$$g(\alpha, \kappa x/k) = 4 C_p \rho_F^{(2\nu-1)} \iint \varkappa^{-(2\nu+1)} \sin^2\left(\varkappa \cdot \alpha/(2\rho_F)\right)$$

$$\times \sin^2\left(\varkappa \cdot \kappa \rho_F/2\right) \frac{d\varkappa}{(2\pi)^2}, \qquad (3.84)$$

where

$$\rho_F = \sqrt{x/k} \qquad (3.85)$$
$$C_p = k^2 l_p C_s. \qquad (3.86)$$

The Fresnel scale, ρ_F, occurs naturally when establishing scale-free units.

3.4.2.1   Weak Scatter   Because the parameter $C_p \rho_F^{(2\nu-1)}$ scales the exponential in (3.82), it can be used to establish the weak-scatter limit. It is readily shown that

$$\lim_{C_p \rho_F^{(2\nu-1)} \to 0} \Phi_I(\kappa) = 4 C_p \varkappa^{-(2\nu+1)} \sin^2\left(\kappa^2 \rho_F^2/2\right). \qquad (3.87)$$

Moreover,

$$\begin{aligned} SI^2 &= 4 C_p \int_0^\infty \varkappa^{-2\nu} \sin^2\left(\kappa^2 \rho_F^2/2\right) \frac{d\kappa}{2\pi} \\ &= 4 C_p \rho_F^{(2\nu-1)} \frac{1}{2^{\nu-1/2}} \int_0^\infty u^{-2\nu} \sin^2\left(u^2\right) \frac{du}{2\pi} \\ &= C_p \rho_F^{(2\nu-1)} \frac{\Gamma\left((2.5 - \nu)/2\right)}{2^{\nu+1/2}\sqrt{\pi}\Gamma\left((\nu + 0.5)/2\right)(\nu - 0.5)}, \end{aligned} \qquad (3.88)$$

which is valid for $0.5 < \nu < 2.5$. The small slope constraint occurs because, in the absence of an inner scale cutoff, high-frequency energy cannot accumulate faster than q^{-1}. At low spatial wavenumbers, the Fresnel-filtering term

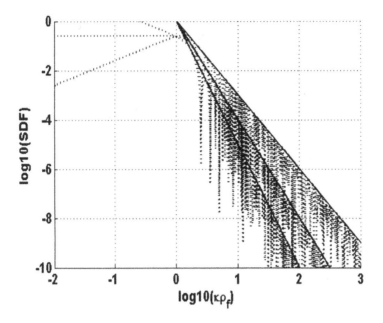

Figure 3.4   Weak scatter limiting form of intensity SDF for different power-law index parameter values.

$\sin^2\left(\kappa^2\rho_F^2/2\right)$ strongly suppresses the low-frequency content. As long as the SDF approaches infinity no faster than q^{-7}, the integral remains finite.[11]

The effect of Fresnel filtering is illustrated in Figure 3.4, which shows a plot of the normalized intensity SDF against the wavenumber normalized to the Fresnel scale. The power-law slopes correspond to $\nu = 1$, 1.5, and 2. For small-sloped spectra, Fresnel filtering strongly suppresses the structure at spatial wavelengths longer than the Fresnel radius. The low-frequency suppression is diminished as the power-law index increases. It will be shown that the intensity spectrum can depart significantly from the weak-scatter theory predictions. However, as a rule, power-law segments generally reflect the integrated phase SDF.

3.4.2.2 Strong Scatter   The strong-scatter limit is much more complicated, but the analysis is simplified significantly in the scale-free limit by using an analytic representation of $g(\boldsymbol{\xi}, \boldsymbol{\eta})$ derived by Rumsey in [48]. The result relies on the structure of the integrand in (3.82). The limiting form is summarized here as

$$g(\boldsymbol{\xi}, \boldsymbol{\eta}) = C(\nu)h(\boldsymbol{\xi}, \boldsymbol{\eta}),$$

[11] $\int \kappa^{n-1}\exp\{-\mu\kappa\}d\kappa = \mu^{-n}\Gamma(n)$ is finite for $n > 0$.

where

$$h(\boldsymbol{\xi},\boldsymbol{\eta}) = \begin{cases} \pm 2\,|\boldsymbol{\xi}|^{2\nu-1} \pm 2\,|\boldsymbol{\eta}|^{2\nu-1} \mp |\boldsymbol{\xi}-\boldsymbol{\eta}|^{2\nu-1} \mp |\boldsymbol{\xi}+\boldsymbol{\eta}|^{2\nu-1} \\ \qquad\quad -\,|\boldsymbol{\xi}|^2 \log |\boldsymbol{\xi}| - |\boldsymbol{\eta}|^2 \log |\boldsymbol{\eta}| \quad\quad \text{for } \nu \neq 1.5 \\ -\tfrac{1}{2}\,|\boldsymbol{\xi}-\boldsymbol{\eta}|^2 \log |\boldsymbol{\xi}-\boldsymbol{\eta}| - \tfrac{1}{2}\,|\boldsymbol{\xi}+\boldsymbol{\eta}|^2 \log |\boldsymbol{\xi}+\boldsymbol{\eta}| \quad \text{for } \nu = 1.5 \end{cases}$$

$$(3.89)$$

$$C(\nu) = \begin{cases} \dfrac{C_p}{2\pi}\dfrac{\Gamma(1.5-\nu)}{\Gamma(0.5+\nu)(2\nu-1)2^{2\nu-1}} & \text{for } 0.5 < \nu < 1.5 \\ \dfrac{C_p}{2\pi} & \text{for } \nu = .5 \\ \dfrac{C_p}{2\pi}\dfrac{\Gamma(2.5-\nu)}{\Gamma(0.5+\nu)(2\nu-1)(2\nu-2)2^{2\nu-1}} & \text{for } 0.5 < \nu < 1.5 \end{cases} , \qquad (3.90)$$

and

$$C_p = k^2 l_p C_s. \qquad (3.91)$$

Although a strictly analytic solution cannot be carried further, it is possible to explore the strong-scatter limiting behavior. To this end, consider the variable change $\alpha = \beta C(\nu)^{-\frac{1}{2\nu-1}}$. Exclusive of $\nu = 1.5$,

$$\Phi_I(\boldsymbol{\kappa}) = C(\nu)^{-\frac{2}{2\nu-1}} \iint \exp\left\{-h\Big(\boldsymbol{\beta},\,\Big[\boldsymbol{\kappa}C(\nu)^{-\frac{1}{2\nu-1}}\Big]C(\nu)^{\frac{2}{2\nu-1}}\rho_F\Big)\right\}$$

$$\exp\left\{-i\Big[\boldsymbol{\kappa}C(\nu)^{-\frac{1}{2\nu-1}}\Big]\cdot\boldsymbol{\beta}\right\}d\boldsymbol{\beta}. \qquad (3.92)$$

To simplify the notation, let

$$\tilde{\Phi}_I(\boldsymbol{\varkappa}) = C(\nu)^{-\frac{2}{2\nu-1}} \iint \exp\left\{-h\big(\boldsymbol{\beta},\,\boldsymbol{\varkappa}C(\nu)^{\frac{2}{2\nu-1}}\rho_F\big)\right\}$$

$$\times \exp\{-i\boldsymbol{\varkappa}\cdot\boldsymbol{\beta}\}d\boldsymbol{\beta}, \qquad (3.93)$$

whereby $\Phi_I(\boldsymbol{\kappa}) = \tilde{\Phi}_I(\boldsymbol{\kappa}C(\nu)^{-\frac{1}{2\nu-1}})$. The strong-scatter limit can be investigated with

$$U(\nu) = C(\nu)^{\frac{2}{2\nu-1}}\rho_F \qquad (3.94)$$

as an expansion parameter. The behavior in the large $U(\nu)$ limit is established by

$$\lim_{\eta\to\infty} h(\boldsymbol{\xi},\boldsymbol{\eta}) = \pm 2\,|\boldsymbol{\xi}|^{2\nu-1}$$

$$+\eta\left(\pm 2\,|\boldsymbol{\eta}/\eta|^{2\nu-1} \mp |\boldsymbol{\xi}/\eta - \boldsymbol{\eta}/\eta|^{2\nu-1}\right.$$

$$\left.\mp |\boldsymbol{\xi}/\eta + \boldsymbol{\eta}/\eta|^{2\nu-1}\right). \qquad (3.95)$$

3.4.2.3 Small-Slope Regime   For $0.5 < \nu < 1.5$, only the leading term in (3.95) needs to be retained, whereby

$$\tilde{\Phi}_I(\varkappa) \simeq C(\nu)^{-\frac{2}{2\nu-1}} \iint \exp\left\{-|\beta|^{2\nu-1}\right\}$$
$$\times \exp\{-i\varkappa \cdot \beta\}d\beta. \tag{3.96}$$

For large $\varkappa$, only small β values contribute. It can be shown by similarity to (3.87) that the same high spatial frequency tail persists.[12] Note also that the $\varkappa$ dependence is contained entirely in the exponential term, which is singular when integrated. It follows that

$$\lim_{U(\nu)\to\infty} \iint \Phi_I(\kappa)\frac{d\kappa}{(2\pi)^2} = 1 \text{ for } 0.5 < \nu < 1.5. \tag{3.97}$$

Thus, in the small-slope regime, which includes Kolmogorov turbulence ($\nu = 4/3$), the scintillation index increases monotonically to unity and the intensity SDF retains the weak-scatter power-law slope at large spatial wavenumbers. This behavior was illustrated in last example of Section 2.2.2. It follows that as $U(\nu) \to \infty$ the limit of $\Gamma_{22}(x, \alpha, 0)$ is equal to $\Gamma_{11}^2(x, \alpha)$.

3.4.2.4 Large-Slope Regime   For $1.5 < \nu < 2.0$, the following approximation can be used:

$$\lim_{\eta\to\infty} h(\xi, \eta) = (2\nu - 1)\left[\xi^2 + (2\nu - 3)(\xi \cdot \eta/\eta)^2\right]\eta^{2\nu-3}, \tag{3.98}$$

whereby

$$\tilde{\Phi}_I(\varkappa)$$
$$\simeq C(\nu)^{-\frac{2}{2\nu-1}} \iint \exp\left\{-(2\nu - 1)\left[\beta^2 + (2\nu - 3)(\beta \cdot \varkappa/\varkappa)^2\right]\right.$$
$$\times (\varkappa U(\nu))^{2\nu-3}\right\} \exp\{-i\varkappa \cdot \beta\}d\beta$$
$$= \frac{\pi}{\sqrt{(2\nu - 2)}\,(2\nu - 1)\,U(\nu)^{2\nu-3}}\varkappa^{3-2\nu}$$
$$\times \exp\left\{-\frac{\varkappa^{5-2\nu}}{4\,(2\nu - 2)\,(2\nu - 1)\,U(\nu)^{2\nu-3}}\right\}. \tag{3.99}$$

One can also show that

$$\lim_{U(\nu)\to\infty} \iint \Phi_I(\kappa)\frac{d\kappa}{(2\pi)^2} = \frac{4\sqrt{2\nu - 2}}{5 - 2\nu} \text{ for } 1.5 < \nu < 2.0. \tag{3.100}$$

[12]See Figure 3.4.

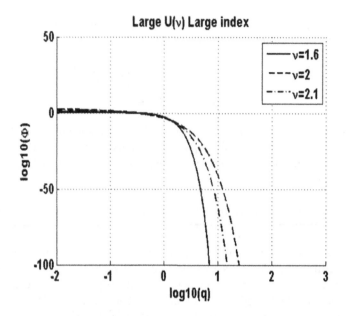

Figure 3.5 Intensity SDF derived from power-law large-slope regime for strong scatter.

Figure 3.5 shows the intensity SDF derived from (3.99). The large spatial-frequency limit does not follow a power law; moreover, the scintillation index in saturation exceeds unity. Figure 3.6 shows a plot of the saturation value versus the spectral index. The full range for valid spectral index values extends to $\nu \sim 2.5$, but the general behavior demonstrated here captures the essential characteristics.

3.4.3 Summary

Although the analytic results presented thus far apply only to the phase screen model, the results capture the essential characteristics of scintillation. Under weak-scatter conditions, the intensity SDF preserves the detail of the underlying SDF as mapped onto the signal phase. This behavior is preserved under strong scatter conditions ($SI \rightarrow 1$) as long as the power-law index is less than $\nu = 1.5$ and $\kappa_f = 2\pi/\rho_F > q_L$. These conditions apply in the small-slope regime, which includes Kolmogorov turbulence ($\nu = 4/3$). In the large-slope regime $\nu > 1.5$ produces strong focusing, as indicated by a local maximum with $SI > 1$. The intensity SDF under strong focusing conditions departs significantly from a simple power-law form.

It is know that exponential models with different power-law regimes might be more appropriate for some applications than an unmodified power law or a power law truncated with an ideal plateau at the outer scale. Indeed,

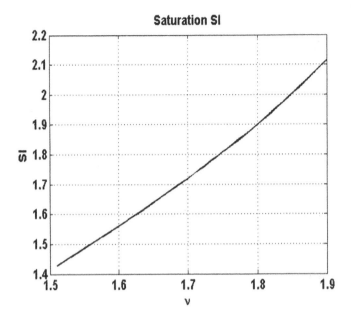

Figure 3.6   Saturation scintillation index for large slope regime in scale-free limit.

Shishov [52] considers a two-component power law as a means to explore the limitations of the unmodified power law. The applicability of an equivalent thin phase screen can also be questioned. The most recent and complete analyses of these topics is contained in the cited works by Martin and Flatte [65], [66] and Flatte and Gerber [71], which will be discussed as numerical simulation examples are presented. However, before doing so it is appropriate to consider higher-order moments and in particular the intensity probability density function (PDF).

3.5 INTENSITY STATISTICS

Because intensity statistics are readily measured, intensity probability distribution functions and the scintillation index were among the first metrics used to characterize scintillation. Intensity statistics also play an important role in evaluating communication system performance. Furthermore, optical systems and microwave imaging systems effectively process two-dimensional fields. Thus, spatial coherence is an important performance metric as well.

   The nature of the weak-scatter model suggested that in-phase and quadrature signal components could be viewed as a jointly Gaussian random variable added to a coherent component ([72] and [20]). However, early studies were influenced by Rytov's method of smooth perturbations [33] and showed that the log-normal distribution gave a better fit to data than the Rice distribution. Indeed, the Rice distribution does not support an SI index greater than unity. For a time the Nakagami distribution [73], which depends only on the SI index, was popular because it predicted SI index values greater than unity. However, the Nakagami distribution has no additional parameters that can vary saturation conditions. Dashen pursued the computation of $\langle I^N \rangle$ using path-integral methods for general power-law structures with some success ([74] and [75]), but a rigorous calculation of the intensity distribution under strong-scatter conditions remains an open challenge.

The most successful models of intensity scintillation are purely phenomenological. The concept that underlies phenomenological modeling is discussed in a concise review by Jakeman [76]. The analysis is based on a model that captures the statistical phenomena in a plausible way. The diffraction model proposed by Jakeman is the two-dimensional random walk

$$\varepsilon(\mathbf{r}, t) = \sum_{n=1}^{N} A_n(\mathbf{r}, t) \exp \{i\varphi_n(\mathbf{r}, t)\}. \tag{3.101}$$

The phase perturbations, $\varphi_n(\mathbf{r}, t)$ are statistically independent and uniformly distributed over 0 to 2π, which restricts the results to strong scatter. The amplitudes, $A_n(\mathbf{r}, t)$, and the number of scattering components, N, are assigned mutually independent statistical distributions. Jakeman summarizes the following two critical properties:

1. If $A_n(\mathbf{r}, t) = A$, the intensity PDF conditioned on knowing A and N is

$$p(I) = \frac{1}{2} \int_0^\infty u J_0(u\sqrt{I}) \left\langle J_0^N(A) \right\rangle_{N,A} du. \tag{3.102}$$

2. If A has the Rayleigh amplitude distribution

$$p(A) = 2A \left\langle A^2 \right\rangle \exp \left\{ -A^2 / \left\langle A^2 \right\rangle \right\}, \tag{3.103}$$

then $p(I) = \exp\{-I/\langle I \rangle\}/\langle I \rangle$, which is the Rayleigh intensity distribution. The same distribution is achieved in the limit for large N. To explore non-Gaussian processes it is necessary to assign a distribution to N or A or both. However, in all cases

$$\langle I^2 \rangle / \langle I \rangle^2 = 2 \langle N(N-1) \rangle / \langle N \rangle^2 + \langle A^4 \rangle / \left(\langle N \rangle \langle A^2 \rangle^2 \right). \tag{3.104}$$

In Figure 2.13 of Chapter 2 local strands of evolving strong intensity were noted. In the abstract model this effect is emulated by a cluster of values of N. On this basis Jakeman proposed the negative binomial distribution

$$p(N) = \left(\begin{array}{c} N + \alpha - 1 \\ N \end{array} \right) \frac{(\langle N \rangle / \alpha)^N}{(1 + \langle N \rangle / \alpha)^{N+\alpha}}. \tag{3.105}$$

For this distribution the normalized second moment of intensity reduces to

$$\langle I^2 \rangle / \langle I \rangle^2 = 2\left(1 + \alpha^{-1}\right) + \langle A^4 \rangle / \left(\langle N \rangle \langle A^2 \rangle^2\right). \qquad (3.106)$$

From this result with $N \to \infty$, the α parameter supports an essentially unlimited departure from the Rayleigh limit $\langle I^2 \rangle / \langle I \rangle^2 = 2$. Thus, a large range of moment-ratio behavior can be matched to theoretical observations. However, the topic remains open to continued research.

3.5.1 Intensity PDFs and Moments

This section summarizes the structure of the intensity PDFs and fractional moments for the distributions that have been used successfully to interpret intensity scintillation. The critical behavior that has captured the most interest is the extreme departures from the exponential distribution that would be expected for underlying Gaussian statistics. The K-distribution has emerged as the most representative of these extreme strong focusing environments.

The K-distribution is defined in terms of the modified Bessel function as

$$p_K(x) = \frac{2}{\Gamma(\alpha)} \left(\alpha x\right)^{(\alpha+1)/2} K_{\alpha-1}\left(2\sqrt{\alpha x}\right). \qquad (3.107)$$

The fractional moments are defined as

$$n^{[m]} = \langle I^m \rangle / \langle I \rangle^m. \qquad (3.108)$$

For the K-distribution it can be show that

$$n_K^{[m]} = \frac{m! \, \Gamma(m + \alpha)}{\alpha^m \Gamma(\alpha)}. \qquad (3.109)$$

Note that $n_K^{[2]} = 2(1 + 1/\alpha)$, which implies that $\alpha = 2/\left(SI^2 - 1\right)$. The singular value at $SI = 1$ corresponds to the exponential distribution. The distribution is, of course, well defined as

$$\lim_{\alpha \to \infty} p_K(x) = \exp\left\{-x\right\}. \qquad (3.110)$$

The form of the log-normal distribution that gives the best fit to weak-to-moderate scintillation distributions is

$$
\begin{aligned}
p_{LN}(x) &= \frac{1}{\sqrt{2\pi\sigma_y^2}\, x} \exp\left\{-\left(\log x + \log\langle I \rangle - \bar{y}\right)^2 / \left(2\sigma_y^2\right)\right\} \\
&= \frac{1}{\sqrt{2\pi\sigma_y^2}\, x} \exp\left\{-\left(\log x + \sigma_y^2/2\right)^2 / \left(2\sigma_y^2\right)\right\} \qquad (3.111)
\end{aligned}
$$

where $x = I/\langle I \rangle$. The logarithmic transformation $\exp\{y\} = I$, where y is Gaussian, establishes the relation

$$\langle I \rangle = \exp\{\langle y \rangle + \sigma_y^2/2\}, \qquad (3.112)$$

which was used to eliminate $\log \langle I \rangle$. By direct computation from (3.111), one can show that

$$\langle I^2 \rangle / \langle I \rangle^2 = \exp\left\{\sigma_y^2\right\}. \tag{3.113}$$

The fractional moments are

$$n_{LN}^{[m]} = (1 + SI)^{m(m-1)/2}. \tag{3.114}$$

For completeness, the Nakagami distribution takes the simplest form

$$p_{NM}(x) = \frac{2M^M (x)^{M-1}}{\Gamma(M)} \exp\left\{-Mx\right\}, \tag{3.115}$$

where $M = 1/SI$.

Figures 3.7, 3.8, and 3.9 show, respectively, the Nakagami, log-normal, and K-distributions for a range of SI spanning weak-to-strong scattering. For the K-distribution, which is mainly applicable to strong scattering, the SI range as been extended and the PDF range plotted in Napiers. The plots show a progression of increasing probability for very large intensity values with increasing SI, corresponding to the strong focusing phenomenon. Figure 3.10 shows the higher fractional moments plotted against the second fractional moment, which have been used extensively for optical data analysis. Data generally follow the K-distribution moments very well. Higher-order intensity moments, if measured accurately, separate the intensity models.

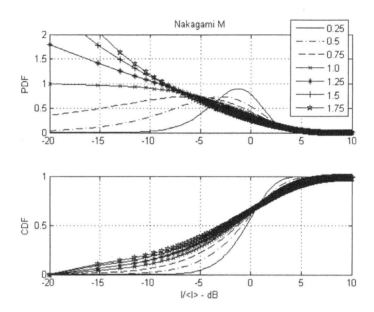

Figure 3.7   PDF and CDF for Nakagami M distribution with varying SI.

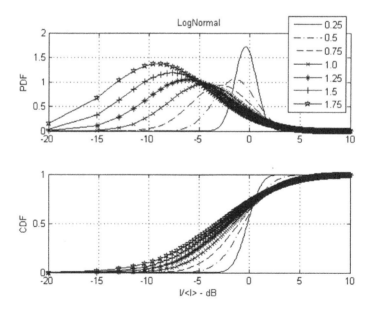

Figure 3.8   PDF and CDF for log-normal distribution with varying SI.

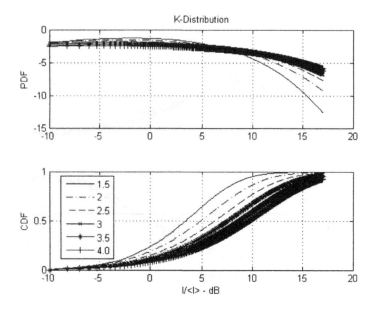

Figure 3.9   PDF and CDF for K-distribution with varying SI.

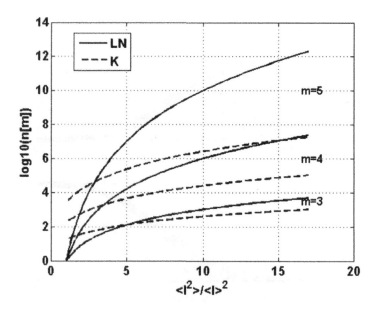

Figure 3.10   Fractional moments for $m = 3, 4$, and 5 plotted against the second fractional moment for the log-normal (solid) and K distributions (dashed).

3.5.2   Simplified Scattering Models

To calculate an appropriate value for the α parameter, Jakeman introduced a model that will be reproduced in a different form to maintain continuity with the development presented in Chapter 2. Consider the scattered field (2.5), which is rewritten here as

$$\mathbf{E}_s(\mathbf{r}) = 2k \int_0^{l_p} \iint \mathbf{E}(\mathbf{r}')S(\mathbf{r}')\frac{\exp\{ik\,|\mathbf{r}-\mathbf{r}'|\}}{2\pi\,|\mathbf{r}-\mathbf{r}'|}d\mathbf{r}'. \qquad (3.116)$$

Let

$$\begin{aligned}
|\mathbf{r}-\mathbf{r}'| &= \sqrt{r^2 - 2\mathbf{r}\cdot\mathbf{r}' + r'^2} \\
&\cong r - \mathbf{r}\cdot\mathbf{r}'/r + r'^2/(2r),
\end{aligned} \qquad (3.117)$$

so that

$$\begin{aligned}
\mathbf{E}_s(\mathbf{r}) \;\cong\; & \frac{k\exp\{-ikr\}}{\pi r} \\
& \times \int_0^{l_p} \iint \mathbf{E}(\mathbf{r}')S(\mathbf{r}') \\
& \times \exp\left\{-ik\left[\mathbf{r}\cdot(\mathbf{r}'/r) - r\,(r'/r)^2/2\right]\right\}d\mathbf{r}'. \qquad (3.118)
\end{aligned}$$

In the phase screen geometry, however, the problem is intrinsically two-dimensional. For example, with plane-wave excitation

$$\begin{aligned}
\mathbf{E}_s(\mathbf{r}) \;\cong\; & \frac{k\exp\{-ikr\}}{\pi r} \\
& \times \mathbf{E}_o \iint \exp\{ikl_p\phi(\varsigma')\} \\
& \times \exp\left\{-i\left[\mathbf{r}\cdot(k\varsigma'/r) - kr\,(\varsigma'/r)^2/2\right]\right\}d\varsigma'. \qquad (3.119)
\end{aligned}$$

The significant range extent of ς' establishes two regimes. If the leading term dominates, the scattered field is the Fourier transform of the phase screen, often referred to as the Fresnel region. As the second term comes into play or dominates, the scattered field enters the strong focusing regime, often referred to as the Fraunhofer region. However, the Fraunhofer is accessible within the bounds of the Fourier transformation region, and it has been explored using the phenomenological model by Michael Berry [77]. His methodology exploits singularities that are manifest in the zero-wavelength limit. The analysis exploits the theory of fractals. The spectral index and the fractal dimension are intimately related. Indeed, Berry's results parallel the regions that showed increasing complexity with increasing spectral index values. This observation will be explored further in the examples below.

3.6 NUMERICAL SIMULATIONS

A direct implementation of the split-step method does not require decorrelation of the structure within an incremental layer. The statistical results based on the Markov approximation, however, are driven by path-integrated structure, which was denoted $\delta\bar{n}(\boldsymbol{\xi})$. The two-dimensional SDF of $\delta\bar{n}(\boldsymbol{\xi})$ is the three-dimensional SDF of $\delta n(x, \boldsymbol{\xi})$ with the κ_x component set equal to zero. A two-dimensional structure realization can be generated directly as

$$
\delta\bar{n}(n\Delta y, m\Delta z) = \sum_{n'=0}^{N} \sum_{m'=0}^{M} \sqrt{\Phi_{\delta\bar{n}}\left(n'\Delta K_y, m'\Delta K_z\right) SF}
$$
$$
\times \varsigma_{n'm'} \exp\left\{i(nn'/N + mm'/M)\right\}, \qquad (3.120)
$$

where ς_{nm} is an uncorrelated, unit-variance sequence of complex Gaussian variates and $\Phi_{\delta\bar{n}}\left(\kappa_y, \kappa_z\right) = \Phi_{\delta n}\left(0, \kappa_y, \kappa_z\right)$. Hermitian symmetry[13] is imposed to ensure that $\delta\bar{n}$ is real. The scale factor

$$
SF = l_p \Delta K_y \Delta K_z / \left(2\pi\right)^2 \qquad (3.121)
$$

is used for convenience. It follows that

$$
\left\langle \frac{1}{NM} \sum_{n=0}^{N} \sum_{m=0}^{M} \delta n^2(n\Delta y, m\Delta z) \right\rangle
$$
$$
= l_p \sum_{n=0}^{N} \sum_{m=0}^{M} \Phi_{\delta n}\left(0, n\Delta K_y, m\Delta K_z\right) dK_y dK_z / \left(2\pi\right)^2 \qquad (3.122)
$$

Thus, the phase perturbation $\delta\phi(\boldsymbol{\xi}) = ik\delta\bar{n}(\boldsymbol{\xi})$ imposes a phase realization that in expectation can be compared directly to theory. Furthermore, a path-integrated deterministic component can be added directly to $\delta\bar{n}(\boldsymbol{\xi})$. The split-step FPE solution itself is completely general.

The three-dimensional split-step recursion defined below is a direct extension of the two-dimensional recursion used for the Chapter 2 examples:

$$
\widehat{\psi}_l(n, m) = \sum_{n'=0}^{N} \sum_{m'=0}^{M} \psi_l(n', m'). * \exp\{ik\delta n(n, m)\Delta x\}
$$
$$
\times \exp\left\{-i(nn'/N + mm'/M)\right\} \qquad (3.123)
$$

$$
\widehat{\psi}_{l+1}(n, m) = \widehat{\psi}_l(n, m). * \exp\left\{ik_x(n, m)l_p\right\} \qquad (3.124)
$$

$$
\psi_{l+1}(n, m) = \frac{1}{NM} \sum_{n'=0}^{N} \sum_{m'=0}^{M} \psi(n', m')
$$
$$
\times \exp\left\{i(nn'/N + mm'/M)\right\}, \qquad (3.125)
$$

---

[13]Hermitian symmetry means $\varsigma_{N-n, M-m} = \varsigma^*_{N-n, M-m}$. It guarantees that δn is real.

where l denotes the layer number, and

$$k_x(n,m) = k\sqrt{1 - \kappa_{n,m}^2/k^2} \tag{3.126}$$

is an N-by-M matrix (in DFT order) that is computed only once. The double summations in (3.123) and in (3.125) define forward and inverse discrete Fourier transformations, respectively. The notation .* means element-by-element multiplication. It is referred to as a Hadamard product. As long as $k_x(n,m)$ is computed in DFT order, no DFT shifts are required to place the zero-frequency term at the center of the array ($n = N/2 + 1$, $m = M/2 + 1$). With a proper choice of sign, the propagation operator $\exp\{ik_x(n,m)l_p\}$ decays exponentially when $k_x > k$, although sampling beyond this limit is unnecessary.

In the theoretical development, scintillation structure was partitioned first into weak- and strong-scatter regimes distinguished by $SI < 1$ and $SI \gtrsim 1$, respectively. Scintillation initiated by an unmodified power-law SDF was further partitioned into small-slope ($\nu < 1.5$) and large-slope ($\nu > 1.5$) regimes. The large-slope regime supports strong focusing as indicated by a local SI maximum greater than unity. In the absence of an outer-scale cutoff, $SI \gtrsim 1$ can persist until the Fresnel scale becomes larger than the largest resolved scale. A two-component power law that starts with a small slope and then transitions to a steeper slope at a prescribed wavenumber is a more realistic representation of real turbulent structures. As discussed earlier in this chapter, at some scale structure must be interpreted as trend-like rather than homogeneous.

3.6.1 Small-Slope Regime

The small-slope scintillation regime has a power-law index parameter in the range $0.5 < \nu < 1.5$. Kolmogorov turbulence is characterized by a power-law radial wavenumber spectrum with $\nu = 4/3$, which is the expected structure for atmospheric turbulence. The first simulation example uses a single 4096-by-4096 realization of $\delta\bar{n}$ with a two-wavelength sampling increment in each direction. At the 1-GHz frequency used for the diffraction computation, the sampled phase screen subtends 2.4 km by 2.4 km. The depth of the propagation space is constrained by the Fresnel radius, ρ_F, which should not exceed the size of the defining phase screen. The curve in Figure 3.11 marked with the * symbols is a computed radial wavenumber SDF from a realization of the phase structure. Because the data are sampled on a rectangular grid, the computed SDF must be interpolated to uniformly sampled polar coordinates prior to integrating over angle to generate the radial SDF estimate. For this computation, a maximum propagation distance of 31.62 km was used. The Fresnel weighting, $\sin^2\left(\kappa^2 x/(2k)\right)$, for the first propagation step (solid) and the last propagation step (dashed) have been superimposed in Figure 3.11 for reference.

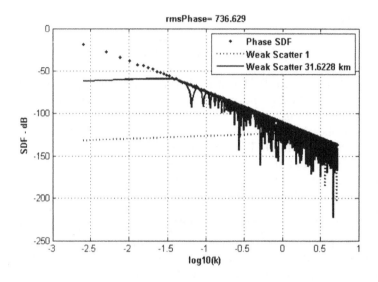

Figure 3.11   Measured phase radial wavenumber SDF with Fresnel filter at first (solid) and last (dashed) propagation steps superimposed.

Although a single propagation step will transform the complex field from the initiating phase screen to the observation plane, it is instructive to follow the evolution of the field structure. Because of the $\rho_F{}^{(2\nu-1)}$ dependence of the weak-scatter theory, the complex field was evaluated over logarithmically spaced planes from the phase screen to the observation plane. The following quantities are computed for each output field: the scintillation index, the 3rd, 4th, and 5th fractional field moments defined by (3.108), and the radial wavenumber SDF for intensity field. The upper frame of Figure 3.12 shows the measured scintillation index versus propagation distance from the phase screen. The lower frame shows the measured 3rd, 4th, and 5th fractional field moments plotted against the second fractional moment, as in Figure 3.10. The fractional moments were computed from the point intensity measurements. The exponential (Rayleigh) distribution has fractional moments

$$\langle I^m \rangle / \langle I \rangle^m = m!,\tag{3.127}$$

which are marked in as pentagrams in Figure 3.12 at the value $FM(2) = 2$, which corresponds to an exponential distribution. Note that the fractional moments are ordered by $\langle I^2 \rangle / \langle I \rangle^2 = SI + 1$, which is not necessarily monotonic with distance.

As predicted by the asymptotic theory for the small-slope regime, the scintillation index increases monotonically to $SI = 1$. Although $SI > 1$ does not occur, the higher fractional moments show that the underlying structure is not strictly exponentially distributed initially. Figure 3.13 shows the evolution with propagation distance (decreasing Fresnel wavenumber) of the measured intensity radial wavenumber SDF. The spectra are qualitatively consistent with the theoretical predictions. The intensity SDF shows the weak-scatter high-frequency power-law tail, and as the scintillation intensity develops, there is a gradual decrease in the local power-law index that would be assigned to the low-spatial frequency region.

Figure 3.14 shows the intensity structure at 31.62 km, which can be compared to Figure 6a in Martin and Flatte [65]. The intensity structure is spatially homogeneous and exhibits some of the characteristics associated with speckle (exponentially distributed intensity). However, the patterned fine structure that can be seen in several subregions is not associated with speckle. Evidently this structure supports the departure of the higher fractional moments from their exponential distribution values.

The mutual coherence function for a plane wave is given formally by (3.32). The theoretical prediction of the mutual coherence function (3.42) is the least mathematically encumbered result. If it is applied exactly, taking the logarithm of the mutual coherence function would isolate the phase structure function. The Fourier transformation of the logarithm of the mutual coherence function would yield the phase SDF, modified at low-spatial wavenumbers by the difference operation in the structure function. Departures from strict homogeneity disrupt this simple theoretical prediction of the mutual coherence function, which limits its practical utility.

Figure 3.15 shows the SDF of the complex field shown in Figure 3.14. The SDF is a more robust measure than the mutual coherence function. The SDF has a bright central region within a broader halo. The fact that the structure approaching the Nyquist spatial sampling frequency is more than 100 dB below the peak intensity confirms that the intensity field is sufficiently sampled. Beyond validating the simulation and establishing a measure of the inverse spatial coherence of the field, however, the complex-field SDF does not yield in situ structure characteristics.

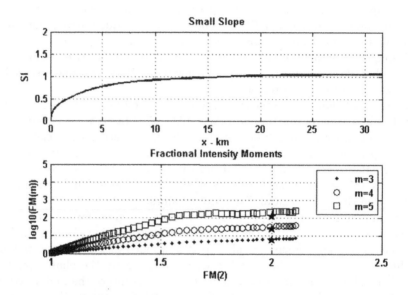

Figure 3.12   Scintillation index (upper frame) and fractional moments (lower frame) plotted against the second fractional moment for $m = 3$, 4, and 5.   The pentagrams at $FM(2) = 2$ are expected values for an exponential distribution.

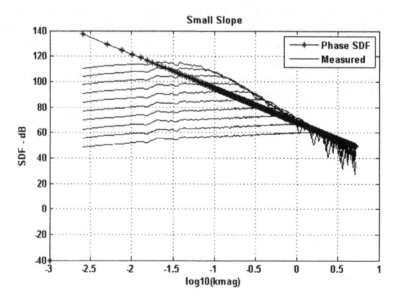

Figure 3.13   Measured radial wavenumber intensity SDFs for $\nu = 2$ realization.

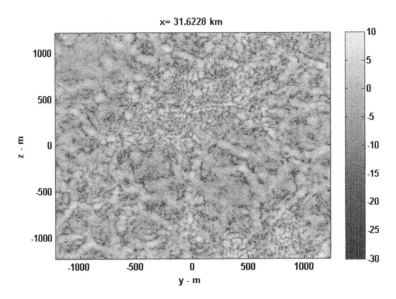

Figure 3.14 Intensity field at 31.62 km from small slope (Kolmogorov) intensity scintillation structure.

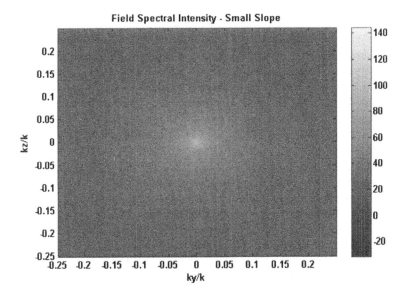

Figure 3.15 Two-dimensional Fourier transformation of complex field whose intensity is shown in Figure 3.14.

3.6.2   Large-Slope Scintillation Regime

The large-slope regime is characterized by strong focusing. Because of the strong sensitivity to changes in the spectral index, asymptotic results were derived only for the subregion $1.5 < \nu < 2.0$. However, two examples are provided at the high end of the $2.0 < \nu < 2.5$ range where scale-free analytic results can be obtained. The sample density was increased from two samples per wavelength to one sample per wavelength to ensure that aliasing would not affect the results. With the same 4096-by-4096 grid, the phase screen size is cut in half. Figure 3.16 shows the scintillation index and the 3rd, 4th, and 5th fractional field moments as in the example in Section 3.6.1 for a power-law realization with $\nu = 2$. As the theory predicts, the SI index reaches a maximum value greater than unity and then decreases toward unity where it remains just above one to 10 km, the maximum range computed. As is consistent with these larger SI values, the fractional moments show large deviations from speckle (the exponential PDF limit). The peak SI value is lower than the predicted saturation value (see Figure 3.6), but $\nu = 2$ is at the limit of the asymptotic range for that approximation. The measured intensity radial wavenumber spectra are shown in Figure 3.17. These results show the expected sharp decay of the intensity radial wavenumber SDF (refer to Figure 3.5), but a high-frequency power-law tail persists. The earlier results presented by Martin and Flatte [65] evidently did not capture the power-law tail because of limited dynamic range supported by the computational resources available.

Figure 3.18 shows the intensity field at 10 km. Here filamentary structures interspersed with fine structure are becoming visible. The filaments are attributed to a self-scaling by random lenses of appropriate strength. Because the measurement is made at a distance beyond the strong-focusing SI peak, the fine structure is evidently a residual of the dissipating large scale structure. Figure 3.19 shows the SDF of the complex field. The weak diagonal bands (more than 100 dB below the peak) may be artifacts of the rectangular computation grid. The main portion of the SDF shows the peak and plateau structure similar to the first example, which verifies the adequacy of the sampling.

For the second large-slope example, a structure realization with $\nu = 2.4$, the scintillation index and the fractional moments are shown in Figure 3.20. The characteristics of the structure evolution are similar to the $\nu = 2$ example, but the strong-focusing SI maximum is larger, as is the departure of the fractional moments from the exponential limit. The radial wavenumber intensity SDFs shown in Figure 3.21 are the SDF at the SI peak and the SDF at 10 km. The SDF at the SI peak is similar to the previous example, but the intensity SDF at 10 km does not capture the expected high frequency rolloff. Finer sampling will capture the peak, but it is the strong focusing regime that is of primare interest here.

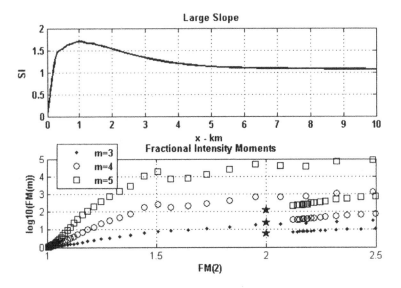

Figure 3.16   Scintillation index (upper frame) and fractional moments (lower frame) plotted against the second fractional moment for $m = 3$, 4, and 5 for a $\nu = 2$ realization.

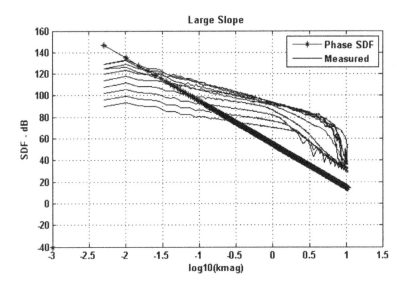

Figure 3.17   Measured radial wavenumber intenstiy SDFs for $\nu = 2$ realization.

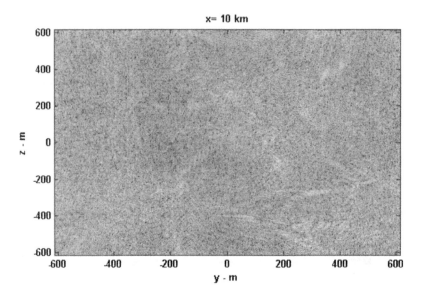

Figure 3.18 Intensity field at 10 km for $\nu = 2$ realization.

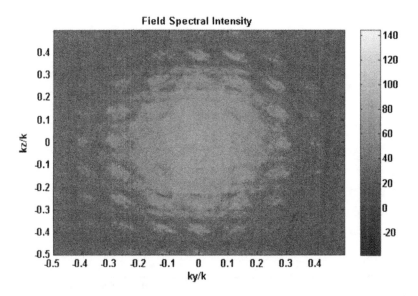

Figure 3.19 SDF of complex field structure for $\nu = 2$ realization shown in Figure 3.18.

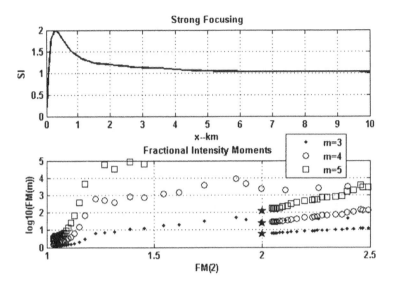

Figure 3.20   Measured SI (upper frame) and fractional moments (lower frame) for the $\nu = 2.4$ realization.

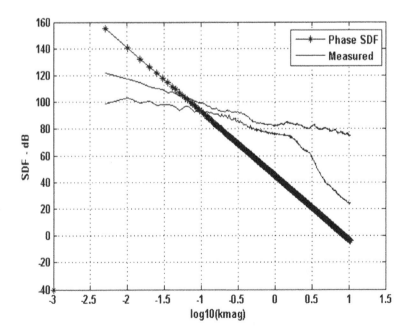

Figure 3.21   Measured radial wavenumber spectra for the $\nu = 2.4$ realization.

Figure 3.22 shows the intensity distribution at the distance where the SI achieves its maximum value near 2. The intensity field shows mainly strong-focusing filaments with one striking pattern of fine strands. This type of scattering is known and has received attention in the literature. A picture similar to Figure 3.22 was used for the cover of the June 1988 issue of *Applied Optics* where the Martin-Flatte article was published. Michael Berry [77] coined the term *diffractal* for this strong-focusing phenomenon. Berry's theoretical analyses use fractal measures to gain insight. The spectral index is closely related to the fractal dimension, which denotes the level of fractal complexity. For completeness, Figure 3.23 shows the SDF of the complex field, which follows the general central peak and halo structure.

Results for extreme scintillation conditions in the scale-free limit build confidence in split-step numerical simulations, which can be implemented with comparative ease. The parameter range that can be explored is too large to summarize here, and departures from the results shown are significant. Cited published results above use numerical simulations extensively as well.

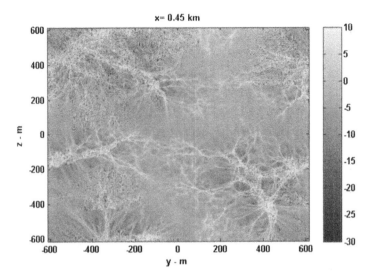

Figure 3.22   Intensity field at SI peak for $\nu = 2.4$ realization.

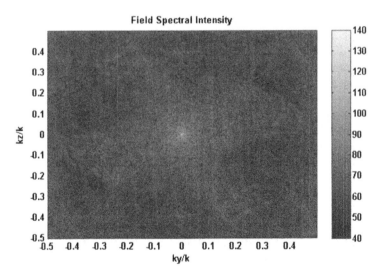

Figure 3.23   Two-dimensional SDF of complex field for $\nu = 2.4$ realization shown in Figure 3.22.

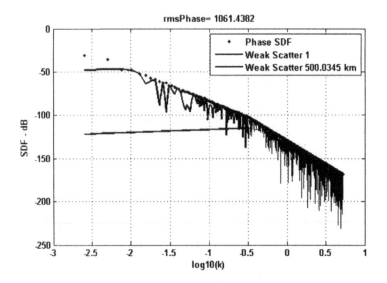

Figure 3.24   Realization of two-component radial wavenumber SDF with weak-scatter theory superimposed at the multilayer exit and 500 km.

3.6.3   Two-Slope Power-Law Scintillation

The final example transitions from a single power-law form to a two-component power-law.   Multiple phase screen implementations have shown that the differences, although significant, are not fundamental.   A two-component power-law form is of particular interest because both in situ ionospheric measurements and satellite scintillation observations support a two-component structure as the source of equatorial ionospheric scintillation ([78] and [79]). However, ionospheric scintillation requires theoretical extensions that will be introduced in Chapter 4.   Numerical simulations have shown that the two-component power-law structure can explain a longstanding ionospheric scintillation puzzle, namely, how to reconcile scintillation levels measured simultaneously at VHF, UHF, and C band [80].

Figure 3.24 shows a realization of a two-component radial wavenumber spectrum otherwise similar to the corresponding spectrum shown Figure 3.11. The free-space simulation was sampled logarithmically, but the phase structure was imparted over five identical 100-m propagation steps.

Because the small-slope segment populates the largest scale sizes, the small-sloped characteristics are expected and seen in the upper frame of Figure 3.25, where SI exceeds unity with a very broad maximum.   The structure evolution shows a hint of strong focusing, whichcould be easily overlooked. The fractional moments shown in the lower frame are similarly transitional.

The measured radial intensity SDFs are shown in Figure 3.26.   The small slope of the spectrum at low radial wavenumbers is difficult to detect because of the Fresnel filtering.   One might easily conclude that the steep slope is the dominant contribution.   The lack of strong focusing is the primary indication that this conclusion is inconsistent with the data.

This was the source of the puzzle alluded to above, but there is more to the story. Ionospheric equatorial irregularities are highly elongated along the Earth's magnetic field lines, which will be accounted for in Chapter 4.   The current example shows that distributing the structure over multiple layers does not change the essential character of the results.   A single layer is adequate to establish representative statistical behavior, although the computational increase for multiple layers is not significant.   The larger challenge is to determine a reasonable path distribution of the structure.

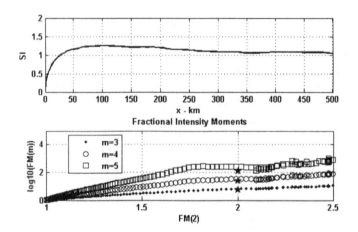

Figure 3.25   Scintillation index and fractional moments for two-component power-law structure.

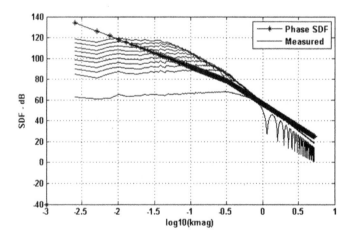

Figure 3.26   Measured radial wavenumber SDF of intensity superimposed on input phase radial wavenumber spectrum.

3.7 STATISTICAL THEORY LIMITATIONS

The *statistical* theory of scintillation is often thought of as the theory of scintillation. Indeed, it seems to embody all the essential characteristics of path-integrated structure development. The statistical theory is viable as long as the formal relation (3.49) applies. In standard developments the critical assumption is the Markov approximation, which effectively ignores structure correlation along the propagation direction. In addition to separating the coupled structure and field correlation measures into a product form, the Markov approximation maps the phase structure from three to two dimensions through integration. The Navikov-Furutsu Theorem sets up the functional relation that absorbs the dimensional change. Operationally, the intensity development over a measure of the media correlation distance must be small. This is appealing because it supports split-step integration of the FPE. However, it is not a necessary condition for split-step integration of the FPE itself. As shown in the example in Section 3.6.2, the FPE supports propagation along a highly elongated structure. The accuracy to which the local focusing is reproduced is open to debate, but the forward approximation requires only that backscatter generated by the in situ structure be small. The main point here is that carefully constructed numerical simulations can set standards against which analytic theory can be tested, not the reverse. Computational electromagnetics comfortably operates with this concept.

In any case, critical insight can be gleaned from supporting numerical simulations without pushing theoretical limits. For example, intensity characteristics initiated by unconstrained large-slope power-law structures tend to develop prominent spatial variations within the confines of a homogeneous structure field that are not revealed by field moment computations. Strong focusing is an extreme example, which has been illuminated by Dashen's theoretical computations [75] and Jakeman's model described in Section 3.5.1. In Chapter 4 local inhomogeneous structure development will not be restricted to the extreme conditions presented in this chapter. The use of numerical simulations will be reinforced in Chapter 5 as the most accurate and efficient means to evaluate the system effects of scintillation.

CHAPTER 4

BEACON SATELLITE SCINTILLATION

Marconi discovered global communication because of reflections from the iono-sphere.

—Umran Ihnan, Stanford University Professor

The launch of the Soviet Satellite Sputnik on October 4, 1957 initiated an era of intensive ionospheric radio propagation research. The Sputnik series and the satellites that followed them carried radio beacons ideally suited for scintillation observation. The beacon transmissions were sensitive mainly to irregularity structures that originate in the ionospheric F-region, from ~ 90 to more than 400 km above the Earth. The physical processes that cause the ionospheric structure have pronounced solar-cycle, seasonal, diurnal, and geographic dependencies. The typical low-altitude orbits with precessing orbital planes provided observations that revealed the morphology of iono-spheric scintillation [2]. It was also recognized that scintillation can impair the performance of satellite communication, navigation, and radar systems. Moreover, the rapidly changing Earth observing geometries present engineer-ing challenges in their own right. These aspects of the satellite scintillation observation will be discussed in Chapter 5.

The Theory of Scintillation with Applications in Remote Sensing, by Charles L. Rino **99**
Copyright © 2011 Institute of Electrical and Electronics Engineers

This chapter extends the FPE theory to accommodate unique aspects of satellite scintillation. The immediate challenge is to incorporate the large extent of the satellite propagation space. The path length from the satellite through the disturbed ionospheric region to the Earth can change by hundreds of kilometers during a typical low-orbiting satellite pass. This implies that the horizontal distance between the origin of the propagation coordinate system in the ionosphere and the ground point of observation spans a comparable distance. To accommodate this changing geometry, a continuously displaced coordinate system centered on the principal propagation direction is used. In a continuously displaced coordinate system, the central analysis region captures the intercept of the principal ray in the plane of observation.

A second challenge is to accommodate the field-aligned anisotropy of iono-spheric irregularities. This anisotropy is a consequence of the fact that charged particles move much more freely along magnetic field lines that across them. As with all tractable analytic results, simplifying assumptions are made. The anisotropy is accommodated by scaling and rotating the princi-pal coordinates that initially characterize isotropic structure. In effect, all structure scales are assumed to have the same anisotropy. The combination of displaced, translated, and scaled coordinates allows all the methods for isotropic irregularities to be applied with appropriate geometric transforma-tions.

The FPE in displaced coordinates is no more difficult to solve than in the normal coordinate system used in Chapters 2 and 3. Moreover, the phase screen structure is effectively a geometric projection of the original isotropic structure that is readily accommodated in simulations. An important aspect of satellite observations is the fact that the observer sees a one-dimensional scan of the diffraction field that is being translated by the combined source, receiver, and media motion as described in Chapter 3. Accommodating this scanning operation involves additional geometric transformations, which are summarized in Appendix A.3. (See Tables 4.1 and 4.2 for symbols and abbreviations.)

Table 4.1 Chapter 4 Symbols

Symbol	Definition
$\psi_{\mathbf{k}}(x, \rho)$	FPE solution in CDCS (4.1)
ρ	Measurement-plane distance in CDCS (4.2)
θ	Propagation angle from x axis
ϕ	Propagation azimuth angle from y axis
$\Theta_{\mathbf{k}}$	Propagation operator in CDCS (4.5)
$\mathbf{k}$	Principal propagation vector (4.3)
$x_P y_P z_P$	Propagation reference coordinate system
$\hat{\mathbf{s}}$	Unit vector along magnetic field in $x_P y_P z_P$ system
ϕ_B, ψ_B	Magnetic azimuth and dip angle in $x_P y_P z_P$ system
$a, b \geq 1$	Principal and secondary anisotropy elongation factors
γ_B	Secondary anisotropy axis orientation angle ($b > 1$)
A, B, C, D	Anisotorpy factors (See Appendix A.2.)
$f(\Delta\rho)$	Measurement plane anisotropy (4.21)
ϑ	Anisotropy scale factor (4.22)
ϖ	Weak scatter SI anisotropy factor (4.39)
$\mathbf{v}_k$	Apparent measurement-plane velocity (4.41)
v_{eff}	Effective velocity for spatial coherence scale
v_{coh}	Effective velocity for temporal coherence scale
SC	Spherical wavefront correction (4.67)

Table 4.2   Chapter 4 Abbreviations

Abbreviation	Definition
CDCS	Continuously displaced coordinate system
MCF	Mutual coherence function
PSD	Power spectral density
GPS	Global Position System
WGS	World Geodetic System
IGRF	International Geomagnetic Reference Field
SGP	Simplified general perturbation
ECI	Earth-centered inertial coordinate system
ECF	Earth-centered fixed coordinate system
TCS	Topocentric coordinate system

4.1   GEOMETRIC CONSIDERATIONS

The scintillation theory development thus far used a reference coordinate system with the principal propagation direction normal to layers in a laterally extended disturbed region. If the principal propagation direction is oblique to the layer, the support space required to capture the evolution of the field as it propagates increases significantly. Over the large propagation distances encountered with typical satellite geometries, the required support space becomes too large for practical analysis. This problem is resolved by introducing a coordinate system with a continuously displaced measurement plane that remains centered on the principal propagation direction. The continuously displaced coordinate system (CDCS) geometry is illustrated in Figure 4.1. A ray aligned with the principal propagation direction intercepts the observation plane at the origin of the $y'z'$ system. This coordinate system was introduced in the seminal paper by Budden [22] as a natural framework to calculate the geometric dependence of scintillation. Briggs and Parkin [19] used an alternate coordinate system oriented normal to the propagation direction irrespective of the layer orientation. To the extent that one is interested only in homogeneous statistical measures, the choice doesn't matter because the final result depends only on the path length within the structure and the propagation direction relative to the magnetic field. For simulations and analyses that address more detailed characteristics of structured media, however, the true layer geometry is essential.

To accommodate propagation along an oblique ray, a product decomposition is introduced as

$$\psi\left(x,\varsigma\right) = \psi_{\mathbf{k}}\left(x,\rho\right)\exp\left\{i\mathbf{k}\cdot\left(x,\varsigma\right)\right\}, \tag{4.1}$$

where ρ is the transverse coordinate in the displaced system,

$$\rho = \varsigma - \tan\theta\widehat{\mathbf{a}}_{k_T}\left(x - x_0\right), \tag{4.2}$$

$\mathbf{k}$ is the fixed propagation vector

$$
\begin{aligned}
\mathbf{k} &= k\left[\sin\theta, \cos\theta\cos\phi, \cos\theta\sin\phi\right] \\
&= \left[kg(\mathbf{k}), \mathbf{k}_T\right],
\end{aligned}
\tag{4.3}
$$

and $\widehat{\mathbf{a}}_{k_T}$ is the unit vector along the transverse component

$$
\begin{aligned}
\widehat{\mathbf{a}}_{k_T} &= \mathbf{k}_T/k_T \\
&= \sec\theta\left[\cos\phi, \sin\phi\right].
\end{aligned}
\tag{4.4}
$$

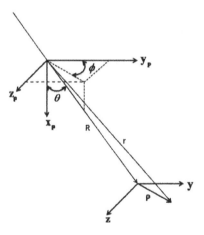

Figure 4.1   Reference coordinate system for oblique propagation relative to disturbed layer with boundaries perpendicular to the x-axis. The orientation of the system is x downward, y eastward, and z southward, .

The propagation angle θ is measured with respect to the propagation axis, not the transverse plane, which is the complement of the usual polar coordinate reference. A generalization of the two-dimensional parabolic wave equation propagator introduced by Costa and Basu [81] follows.

It can be shown by direct computation starting with (4.1) that

$$\psi_{\mathbf{k}}(x;\boldsymbol{\rho}) = \iint \widehat{\psi}_{\mathbf{k}}(x_0;\boldsymbol{\kappa}) \exp\left\{i\left(kg\left(\boldsymbol{\kappa}+\mathbf{k}_T\right)-\tan\theta\widehat{\mathbf{a}}_{k_T}\cdot\boldsymbol{\kappa}\right)\left(x-x_0\right)\right\}$$
$$\times \exp\left\{i\boldsymbol{\rho}\cdot\boldsymbol{\kappa}\right\} \frac{d\boldsymbol{\kappa}'}{\left(2\pi\right)^2}. \tag{4.5}$$

In effect (4.5) defines a new propagation operator, $\Theta_{\mathbf{k}}$, that can be applied to $\psi_{\mathbf{k}}(x,\boldsymbol{\rho})$ in the continuously displaced coordinate system. The FPE is rewritten in terms of $\psi_{\mathbf{k}}(x,\boldsymbol{\rho})$ and $\Theta_{\mathbf{k}}$ as

$$\frac{\partial \psi_{\mathbf{k}}(x,\boldsymbol{\rho})}{\partial s} = ik\Theta_{\mathbf{k}}\psi(x,\boldsymbol{\rho}) + ik\delta n(x,\boldsymbol{\rho})\psi_{\mathbf{k}}(x,\boldsymbol{\rho}), \tag{4.6}$$

where

$$\frac{\partial \psi_{\mathbf{k}}(x,\boldsymbol{\rho})}{\partial s} = \lim_{\Delta x \to 0} \frac{\psi_{\mathbf{k}}(x+\Delta x,\boldsymbol{\rho}+\tan\theta\widehat{\mathbf{a}}_{k_T}\Delta x) - \psi_{\mathbf{k}}(x,\boldsymbol{\rho})}{\sec\theta\Delta x} \tag{4.7}$$

defines the directional derivative along $\widehat{\mathbf{a}}_k$. In the absence of diffraction the phase perturbation in the continuously displaced system is

$$\delta\phi_{\mathbf{k}}(l_p,\boldsymbol{\rho}) = k\sec\theta \int_0^{l_p} \delta n(\eta,\boldsymbol{\rho}+\tan\theta\widehat{\mathbf{a}}_{k_T}\eta)d\eta. \tag{4.8}$$

A split-step solution to (4.6) can be implemented as before by generating a phase perturbation defined by (4.8) and then propagating the field forward using (4.5).

4.2 PHASE STRUCTURE REVISITED

Assuming as before that the refractive index structure can be characterized by a homogeneous random process, path-integrated phase remains as a sufficient representation of the in situ structure for FPE analysis. The representations developed in Section 3.1.5 need only be translated to the CDCS. The integrated phase ACF follows directly from (4.8) as

$$R_{\delta\phi_{k}}(\Delta\boldsymbol{\rho}) = k^2 l_p \sec^2\theta \int_{-l_p}^{l_p} \left(1 - \frac{|\Delta\eta|}{l_p}\right) R_{\delta n}\left(\Delta\eta,\Delta\boldsymbol{\rho}-\tan\theta\widehat{\mathbf{a}}_{k_T}\Delta\eta\right) d\Delta\eta$$
$$\simeq k^2 l_p \sec^2\theta \int_{-\infty}^{\infty} R_{\delta n}\left(\Delta\eta,\Delta\boldsymbol{\rho}-\tan\theta\widehat{\mathbf{a}}_{k_T}\Delta\eta\right) d\Delta\eta. \tag{4.9}$$

The two-dimensional phase SDF, which is obtained by direct computation, is

$$\Phi_{\delta\phi_k}(\boldsymbol{\kappa}) = \iint R_{\delta\phi_k}(\Delta\rho) \exp\{-i\boldsymbol{\kappa}\cdot\Delta\rho\} d\Delta\rho$$
$$= k^2 l_p \sec^2\theta \Phi_{\delta n}(\tan\theta\boldsymbol{\kappa} \cdot \hat{\mathbf{a}}_{k_T}, \boldsymbol{\kappa}). \qquad (4.10)$$

The translation of phase structure to the CDCS is achieved simply by replacing the κ_x wavenumber of the three-dimensional SDF in the normal system by $\tan\theta\boldsymbol{\kappa} \cdot \hat{\mathbf{a}}_{k_T}$ and scaling the translated SDF by $\sec^2\theta$. Recall that for normal propagation, the κ_x wavenumber is equal to zero, which follows from (4.5) with $\theta = 0$.

4.2.1   Anisotropy

Following Singleton's approach [43], an anisotropic correlation function is generated by a series of rotations followed by a scaling of the displacement coordinates that characterize the isotropic correlation function. In an earlier development [24], a coordinate system with its principal axis of anisotropy contained in the vertical reference plane was used. However, because magnetic field vectors are reported in topocentric coordinates it is more efficient to adopt the topocentric system at the outset. To maintain consistency with the notation of Chapters 1, 2, and 3, the coordinate system shown in Figure 4.1 is employed with positive z_P southward, positive y_P eastward, and positive x_P downward. The Earth's magnetic field direction in the $x_P y_P z_P$ system is defined as

$$\hat{\mathbf{s}} = [\sin\psi_B, \cos\psi_B \sin\phi_B, \cos\psi_B \cos\phi_B], \qquad (4.11)$$

where ϕ_B defines the orientation of the plane of the magnetic field, and ψ_B is the angle from the horizontal plane (the dip angle). A rotation about the x_P axis through ϕ_B places the principal irregularity axis in the $x'z'$ plane. A rotation about z' through ψ_B aligns the z'' axis with the principal irregularity axis. A final rotation about y'' through γ_B aligns the y''' axis with a second elongation axis if the second elongation factor b is greater than one. The $x'''y'''z'''$ system now coincides with the principal axes of elongation. A scaling of y''' by $1/a$ and z''' by $1/b$ generates a displacement that, when evaluated with an isotropic correlation function, generates $x_P y_P z_P$ contours of constant correlation. The principal elongation axis lies along y''' with the z''' defining the transverse plane. In Singleton's original notation $r = y'''$, $s = z'''$, and $t = x'''$.

The rotations and scaling just described can be written in matrix form as

$$y = \sqrt{\left(\overline{D}_{ab}^{-1}\overline{U}_{\gamma_B}\overline{U}_{\psi_B}\overline{U}_{\phi_B}\Delta\varsigma\right)^T \left(\overline{D}_{ab}^{-1}\overline{U}_{\gamma_B}\overline{U}_{\psi_B}\overline{U}_{\phi_B}\Delta\varsigma\right)}, \qquad (4.12)$$

where $\overline{U}_{\phi_B}$, $\overline{U}_{\psi_B}$, and $\overline{U}_{\gamma_B}$ are rotation matrices defined in Appendix A.3, and the elongation factors are the diagonal elements of the matrix

$$\overline{D}_{ab} = \begin{bmatrix} 1 & 0 & 0 \\ 0 & a & 0 \\ 0 & 0 & b \end{bmatrix}. \tag{4.13}$$

It is convenient to summarize the results in terms of a single matrix $\overline{C}$ and the complementary matrix $\widehat{C}$ that defines constant SDF intensity surfaces. The defining quadratic-function relations follow:

$$y^2 = \Delta\varsigma^T \overline{C} \Delta\varsigma \tag{4.14}$$

$$q^2 = \kappa^T \widehat{C} \kappa \tag{4.15}$$

The elements of the $\widehat{C}$ and $\overline{C}$ matrices are computed in Appendix A.3 as functions of the polar angles ψ_B, ϕ_B, and the orientation angle γ_B together with the elongation factors a and b. Note that if $b = 1$, the γ rotation has no effect.

The general anisotropic SDF under the Singleton model is

$$\Phi_{\delta n}(\kappa) = ab \left\langle \delta n^2 \right\rangle Q \left(\sqrt{\kappa^T \widehat{C} \kappa} \right), \tag{4.16}$$

where Q was defined in Section 3.1.4. The ab scaling preserves the normalization

$$\left\langle \delta n^2 \right\rangle = \iiint \Phi_{\delta n}(\kappa) \frac{d\kappa}{(2\pi)^3}. \tag{4.17}$$

Evaluating (4.15) with $\kappa_x = \tan\theta \kappa' \cdot \widehat{\mathbf{a}}_{k_T}$ generates the two-dimensional quadratic form

$$q^2 = A\kappa_y^2 + B\kappa_y\kappa_z + C\kappa_z^2. \tag{4.18}$$

The coefficients A, B, and C are also defined in Appendix A.2. The SDF of the integrated phase in the CDCS anisotropic system can be written as

$$\begin{aligned} \Phi_{\delta\phi}(\kappa) &= k^2 l_p ab \sec^2\theta \left\langle \delta n^2 \right\rangle Q \left(\tan\theta\kappa \cdot \widehat{\mathbf{a}}_{k_T}, \kappa \right) \\ &\simeq k^2 l_p ab \sec^2\theta C_s \left(A\kappa_y^2 + B\kappa_y\kappa_z + C\kappa_z^2 \right)^{-(2\nu+1)}, \end{aligned} \tag{4.19}$$

where $Q(q)$ is defined in (3.18). Upon applying a coordinate rotation and scaling, the integrand argument Q can be made isotropic. This yields the anisotropic path-integrated phase ACF

$$\begin{aligned} R_{\delta\phi}(\Delta\rho) &= k^2 l_p ab \sec^2\theta \left\langle \delta n^2 \right\rangle \\ &\quad \times \iint Q \left(A\kappa_y^2 + B\kappa_y\kappa_z + C\kappa_z^2 \right) \exp\{i\kappa\cdot\Delta\rho\} \frac{d\kappa}{(2\pi)^2} \\ &= k^2 \left\langle \delta n^2 \right\rangle l_p \vartheta \varkappa \sec\theta \Re(f(\Delta\rho)), \end{aligned} \tag{4.20}$$

where $\Re(y)$ and $\varkappa$ are defined in Chapter 3, and

$$f(\Delta\rho) = \left[\frac{C\rho_y^2 - B\rho_y\rho_z + A\rho_z^2}{(AC - B^2/4)}\right]^{1/2} \tag{4.21}$$

$$\vartheta = \frac{ab}{\sqrt{(AC - B^2/4)}\cos\theta}. \tag{4.22}$$

A comparison to (4.19) and (4.20) shows that the respective ellipses are orthogonal, as the inverse relation between spatial wavenumber and structure scale predicts. The function $f(\Delta\rho)$ defines measurement plane elliptical contours of constant correlation that are orthogonal to the contours of constant integrated phase spectral intensity. Note that A, B, and C refer to the integrated phase spatial frequency domain quadratic form (4.18).

These results extend the analytic structure models developed in Chapter 3 to accommodate both anisotropy and the CDCS. Contours of constant spatial correlation of the integrated phase are shadow-like projections of the ellipsoids of constant refractive index correlation. The full range of anisotropy in the Singleton model involves two elongation factors, a and b. When $b > 1$, the additional angle γ_B that defines the orientation of a second elongation axis must be specified. However, it has been recognized since the early development of scintillation theory that the principal geometric dependence imposed by anisotropy depends on the Briggs-Parkin angle [19], which is defined by the dot product

$$\cos\psi_{BP} = \widehat{\mathbf{a}}_k \cdot \widehat{\mathbf{s}}. \tag{4.23}$$

4.2.2 Weak Scatter

In the CDCS the weak-scatter intensity SDF takes the form

$$\begin{aligned}\Phi_I(\boldsymbol{\kappa}) &= 4\Phi_{\delta\phi}(\boldsymbol{\kappa})\sin^2(k\left(g(\boldsymbol{\kappa}+\mathbf{k}_T) - (\tan\theta)\,\widehat{\mathbf{a}}_{k_T}\cdot\boldsymbol{\kappa}\right)x) \\ &\simeq 4\Phi_{\delta\phi}(\boldsymbol{\kappa})\sin^2\left(\left(\kappa^2 + \tan^2\theta\,(\widehat{\mathbf{a}}_{k_T}\cdot\boldsymbol{\kappa})^2\right)\frac{x\sec\theta}{2k}\right).\end{aligned} \tag{4.24}$$

The simplification in the second line is obtained by using the following narrow-angle scatter approximation

$$\begin{aligned}kg(\boldsymbol{\kappa}+\mathbf{k}_T) &= k\cos\theta \\ &\times\left[1 - \left(1 - \left(\frac{\kappa}{k\cos\theta}\right)^2 - 2\frac{\tan\theta\widehat{\mathbf{a}}_{k_T}\cdot\boldsymbol{\kappa}}{k\cos\theta}\right)^{1/2}\right] \\ &\simeq (\tan\theta)\,\widehat{\mathbf{a}}_{k_T}\cdot\boldsymbol{\kappa} + \frac{\kappa^2 + \tan^2\theta\,(\widehat{\mathbf{a}}_{k_T}\cdot\boldsymbol{\kappa})^2}{2k\cos\theta}.\end{aligned} \tag{4.25}$$

The scintillation index, obtained as before by integrating $\Phi_I(\boldsymbol{\kappa})$ over $\boldsymbol{\kappa}$ and using the narrow-angle scatter approximation, is given by

$$SI^2 \;\simeq\; 4k^2 l_p ab \sec^2\theta \left\langle \delta n^2 \right\rangle \iint Q(A\kappa_y^2 + B\kappa_y\kappa_z + C\kappa_z^2)$$
$$\times \sin^2\left(\left(\kappa^2 + \tan^2\theta\,(\widehat{\mathbf{a}}_{k_T}\!\cdot\!\boldsymbol{\kappa})^2\right) \frac{x\sec\theta}{2k} \right) \frac{d\boldsymbol{\kappa}}{(2\pi)^2}. \tag{4.26}$$

The $\widehat{\mathbf{a}}_{k_T}\!\cdot\!\boldsymbol{\kappa}$ term can be eliminated by rotating $\boldsymbol{\kappa}$ through the propagation azimuth angle ϕ and scaling the rotated coordinate by $\sec\theta$ (See Appendix A.2). The result is

$$SI^2 \;\simeq\; 4k^2 l_p ab \sec\theta \left\langle \delta n^2 \right\rangle \iint Q(A'\kappa_y^2 + B'\kappa_y\kappa_z + C'\kappa_z^2)$$
$$\times \sin^2\left(\kappa'^2 \rho_F'/2 \right) \frac{d\boldsymbol{\kappa}}{(2\pi)^2}, \tag{4.27}$$

where

$$A' \;=\; \left[A\cos^2\phi + B\sin\phi\cos\phi + C\sin^2\phi\right]\cos^2\theta \tag{4.28}$$
$$B' \;=\; \left[B\cos 2\phi + (C-A)\sin 2\phi\right]\cos\theta \tag{4.29}$$
$$C' \;=\; A\sin^2\phi - B\sin\phi\cos\phi + C\sin^2\phi \tag{4.30}$$

The Fresnel radius in the CDCS becomes

$$\rho_F' = \frac{x\sec\theta}{k}. \tag{4.31}$$

Now consider the scale-free power-law form of the structure SDF,

$$\left\langle \delta n^2 \right\rangle Q(A'\kappa_y^2 + B'\kappa_y\kappa_z + C'\kappa_z^2) \simeq C_s \left(A'\kappa_y^2 + B'\kappa_y\kappa_z + C'\kappa_z^2\right)^{-(\nu+1/2)}. \tag{4.32}$$

With some straightforward additional variable changes it follows that

$$SI^2 \;=\; 4k^2 l_p ab \sec\theta\, C_s \rho_F^{(2\nu-1)} 2^{\nu-1/2} \frac{1}{2\pi} \int_0^\infty q^{-2\nu}\sin^2\left(q^2\right) dq$$
$$\times \begin{cases} \frac{2}{\pi}\int_0^{\pi/2}\left(A'' - (A''-C'')\sin^2\phi\right)^{-(\nu+1/2)} d\phi \\ \frac{2}{\pi}\int_0^{\pi/2}\left(C'' - (C''-A'')\cos^2\phi\right)^{-(\nu+1/2)} d\phi \end{cases}, \tag{4.33}$$

where

$$A'' \;=\; \left(A' + C' - D'\right)/2 \tag{4.34}$$
$$C'' \;=\; \left(A' + C' - D'\right)/2 \tag{4.35}$$

and

$$D' = \sqrt{(C' - A') - B'^2} \qquad (4.36)$$

The integral over q was identified in Chapter 3 as

$$\int_0^\infty q^{-2\nu} \sin^2\left(q^2\right) dq = \frac{\sqrt{\pi}}{2} \frac{\Gamma\left((2.5 - \nu)/2\right)}{\Gamma\left((\nu + 0.5)/2\right)(\nu - 0.5)}. \qquad (4.37)$$

The integral over ϕ can be evaluated in terms of the Gauss hypergeometric function $F(\alpha, \beta; \gamma; \varsigma)$. It follows that

$$SI^2 = C_p \sec\theta \rho_F^{(2\nu-1)} \frac{\Gamma\left((2.5 - \nu)/2\right)}{2^{\nu+1/2}\sqrt{\pi}\Gamma\left((\nu + 0.5)/2\right)(\nu - 0.5)}\varpi, \qquad (4.38)$$

where phase turbulent strength $C_p = k^2 l_p C_s$ was defined earlier, and

$$\varpi = \begin{cases} abF\left(\nu + 1/2, 1/2; 1; \frac{A''-C''}{A''}\right) A''^{-(\nu+1/2)} & \text{if } A'' > C'' \\ abF\left(\nu + 1/2, 1/2; 1; \frac{C''-A''}{C''}\right) C''^{-(\nu+1/2)} & \text{if } C'' > A'' \end{cases}. \qquad (4.39)$$

If $A'' = C''$, then $\varpi = 1$, which is the isotropic limit. The term ϖ is a correction to the result for isotropic structure in the CDCS. The model approaches the two-dimensional limiting form when A'' and C'' are nearly equal.

Figure 4.2 shows a meridian slice of the SI variation for an isotropic layer and elongation 10:1 at $\theta = 45$ degrees. There is a 10:1 SI enhancement where the propagation path aligns itself with the axis of elongation or where $\psi_{BP} = 0$. For isotropic irregularities, the SI index varies with the path length intercepting the layer. Ionospheric observations of low-orbiting satellites in the auroral zone have produced enhancements that confirm the expected geomagnetic field alignment of the structures [82].

4.3   COMPLEX FIELD COHERENCE REVISITED

The implicit phase reference for coherent field measurement is the point at which a ray along the principal propagation direction intercepts the measurement plane ($\rho = 0$). The practical means of providing such a phase reference will be discussed in Chapter 5. Assuming that plane-wave excitation is used to generate the solution to the FPE in the CDCS, the measurement plane field is $\psi_k(x, \rho; t)$. However, to proceed, the sources of temporal variation must be identified explicitly. The simplest model assumes that the media structure is invariant over the measurement interval. In that case, the observed time variation comes from structure drift and the apparent motion of the origin of coordinates induced by the source and receiver motion. The spatial structure of the field is translated in time with the apparent velocity $\mathbf{v_k}$ as follows:

$$\rho \leftarrow \rho + \mathbf{v_k}t. \qquad (4.40)$$

From (4.2) the effective velocity in the CDCS is

$$\mathbf{v}_k = \left(\mathbf{v}_T - (\tan\theta)\,\widehat{\mathbf{a}}_{k_T} v_x\right), \tag{4.41}$$

where

$$\mathbf{v}_T = \mathbf{v}_d - \mathbf{v}_p. \tag{4.42}$$

The velocity $\mathbf{v}_d$ is the structure drift velocity and $\mathbf{v}_p$ is the reference source-induced motion of the coordinate system. Note that in the CDCS, vertical displacements shift the coordinate system and thereby give rise to an apparent translation.

The x variable changes with time as well, but the change induced by moderate changes along the propagation reference axis is usually small. To confirm this assumption, the change in SI due to parameter variation during a single measurement should be negligible. There is an important exception to slowly varying change along the propagation direction. Indeed, early rocket-borne beacon experiments presented geometries in which the path velocity component dominated the evolving structure [83]. Most rocket observations fall into this regime. The reference system must be recast to accommodate the evolving layer thickness to properly interpret such data [78].

4.3.1 Space Time Mutual Coherence

Measurements of $\psi_k\left(x;\rho,t\right)$ with separate antennas provide the most direct sampling of the complex field structure. To the extent that height differences between the receivers are small, which can be checked as noted above

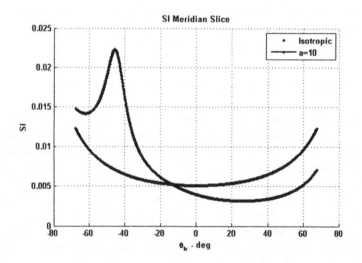

Figure 4.2   Meridian plane variation of SI for isotropic structure (heavy dashed) and structures with 10:1 elongation at a 45 degree declination angle (solid).

by comparing the SI index measured at each antenna, a direct measure of $\Gamma_{11}(x;\varsigma_1,\varsigma_1')$ as defined by (3.50) can be constructed. In the continuously displaced coordinate system the mutual coherence function takes the simple form

$$
\begin{aligned}
R_{\psi_{\mathbf{k}}}(\Delta\boldsymbol{\rho} + \mathbf{v}_k\Delta t) &= \langle \psi_{\mathbf{k}}(x; \boldsymbol{\rho} + \mathbf{v}_k t)\, \psi_{\mathbf{k}}^*(x; \boldsymbol{\rho}' + \mathbf{v}_k t')\rangle \\
&= \exp\{-D_{\delta\phi}(f(\Delta\boldsymbol{\rho} + \mathbf{v}_k\Delta t))\}.
\end{aligned} \tag{4.43}
$$

Applying the spectral model and the anisotropy translations explicitly gives the phase structure function as

$$
D_{\delta\phi}(f(\Delta\boldsymbol{\rho} + \mathbf{v}_k\Delta t)) = k^2\vartheta\,\langle\delta n^2\rangle\, l_p \sec\theta\,(1 - \varkappa\Re(f(\Delta\boldsymbol{\rho} + \mathbf{v}_k\Delta t))). \tag{4.44}
$$

By using the scale-free power-law limit the phase structure function is simplified further to

$$
D_{\delta\phi}(f(\Delta\boldsymbol{\rho})) \simeq C_{\delta\phi}\vartheta \sec\theta f(\Delta\boldsymbol{\rho} + \mathbf{v}_k\Delta t)^{2\nu-1}, \tag{4.45}
$$

where $C_{\delta\phi}$ is the phase structure constant defined by (3.41). Note that geometry and the anisotropy affect spatial coherence in two ways. They change the effective scale through the geometric dependence of the argument of f and through the geometric factors that modulate the phase structure constant, namely $\vartheta \sec\theta$.

Measurements of spaced-receiver correlation functions from each of $N_{\mathrm{rec}}!/2$ receiver pairs provide the following family of time-series measurements

$$
d_{nm}(\Delta t) = R_{\psi_{\mathbf{k}}}(\Delta\boldsymbol{\rho}_{mn} + \mathbf{v}_k\Delta t). \tag{4.46}
$$

The measurements can be processed to extract estimates of $\mathbf{v}_k$ anisotropy and layer height [84]. The methodology has been known for decades [85], and it continues to be used for routine drift-velocity and structure-height estimates [86].

4.3.2   Time Series Measurement

Single receiver measurements of the complex field are recorded as time series. If the structure is statistically homogeneous, the temporal autocorrelation function can be written as

$$
\begin{aligned}
R_{\psi_{\mathbf{k}}}(\mathbf{v}_k\Delta t) &= \langle \psi_{\mathbf{k}}(x; \mathbf{v}_k t)\, \psi_{\mathbf{k}}^*(x; \mathbf{v}_k t')\rangle \\
&= \exp\{-D_{\delta\phi}(v_{\mathrm{eff}}\Delta t)\},
\end{aligned} \tag{4.47}
$$

where the following effective velocity scales the time displacement:

$$
\begin{aligned}
v_{\mathrm{eff}} &= f(\mathbf{v}_k) \\
&= \left[\frac{Cv_{ky}^2 - Bv_{ky}v_{kz} + Av_{kz}^2}{AC - B^2/4}\right]^{1/2}.
\end{aligned} \tag{4.48}
$$

From (4.45), it follows that

$$R_{\psi_k}(\mathbf{v}_k \Delta t) = \exp\left\{-C_{\delta\phi} v_{\text{coh}} |\Delta t|^{2\nu-1}\right\}, \qquad (4.49)$$

where

$$v_{\text{coh}} = \vartheta v_{\text{eff}} \sec\theta. \qquad (4.50)$$

These effective velocity definitions combine several geometric factors that influence mutual coherence measurement.

The formal power spectral density (PSD) of the time series $\psi_k(x; \mathbf{v}_k t)$ is, in expectation, the Fourier transform of the mutual coherence function (MCF) R_{ψ_k} along the effective scan direction

$$\varphi(f) = \int R_{\psi_k}(\mathbf{v}_k \Delta t) \exp\left\{-2\pi i f \Delta t\right\} d\Delta t, \qquad (4.51)$$

but the result admits no analytic solution and correlation measurements themselves are not readily manipulated in the manner implied by their error-free expectation. For example, it follows from (4.47) that the mutual coherence function is positive definite. Measurements show a persistent negative tail under weak-scatter conditions that is attributed to departures from statistical homogeneity [87]. A simpler and potentially more robust interpretation of complex data is suggested by the one-dimensional PSD of the integrated phase. The modified power-law form of the integrated phase ACF admits the analytic solution

$$\begin{aligned} \varphi(f) &= \int R_{\delta\phi}(v_{\text{eff}} \Delta t) \exp\left\{-i f \Delta t\right\} d\Delta t \\ &= \frac{T}{[f_L^2 + f^2]^\nu}, \end{aligned} \qquad (4.52)$$

where

$$f_L = v_{\text{eff}} q_L / (2\pi), \qquad (4.53)$$

and

$$T = v_{\text{eff}} \vartheta \sec\theta C_p \frac{\sqrt{\pi}\Gamma(\nu)}{(2\pi)^{2\nu+1}\Gamma(\nu+1/2)}. \qquad (4.54)$$

Because the geometric factors that modulate T are nominally time invariant, v_{eff} is the proper temporal-to-spatial conversion factor.

The integrated phase spectrum is not the phase of the complex signal, but experience and simulations suggest that the envelope of the phase spectrum coincides with the spectrum of the path-integrated phase. Moreover, the high-frequency tail of the intensity spectrum has the same envelope under moderate to strong scattering conditions. The fact that the power-law index of the one-dimensional spectrum is one less than the three-dimensional power-law index seems to be robust, but different spectral hypotheses should be investigated on a case-by-case basis.

4.3.3 Frequency Coherence

Most scintillation analyses and measurements assume coherence over the frequency band. A measure of frequency coherence is necessary to verify this assumption. If the signal bandwidth is large enough, statistical decorrelation can degrade the integrity of the waveform occupying the band of frequencies. The two-frequency coherence function is the most widely used measure of such effects. The result from Chapter 3.3, (3.62), is reproduced here for reference as

$$
\begin{aligned}
\Gamma_{11}(x; \Delta\varsigma; \quad k_1, k_2) = {}& \exp\left\{-l_p R_{\delta\overline{n}}(0)(k_1 - k_2)^2/2\right\} \\
& \times \iint \exp\left\{-l_p D_{\delta\overline{n}}(\Delta\varsigma)k_1 k_2\right\} \\
& \times \iint \exp\left\{-i\left(k_1 g_1(\kappa) - k_2 g_2(\kappa)\right)x\right\} \\
& \times \exp\left\{-i\kappa \cdot (\Delta\varsigma' - \Delta\varsigma)\right\}\frac{d\kappa}{(2\pi)^2}d\Delta\varsigma'.
\end{aligned}
\tag{4.55}
$$

This relation is translated to the CDCS by the replacements $\Delta\varsigma \leftarrow \Delta\rho$ and $g(\kappa) \leftarrow g(\tan\theta \widehat{\mathbf{a}}_{k_T} \cdot \kappa, \kappa_T)$. Using the parabolic approximation (4.25) as before, the linear terms cancel and the propagation term simplifies to

$$
(k_1 g_1(\kappa) - k_2 g_2(\kappa)) \simeq \frac{\kappa^2 + \tan^2\theta\left(\widehat{\mathbf{a}}_{k_T}\cdot\kappa\right)}{2\cos\theta}\left(\frac{1}{k_1} - \frac{1}{k_2}\right).
\tag{4.56}
$$

The CDCS forms of the structure function terms are

$$
kl_p R_{\delta\overline{n}}(0) = k^2\vartheta\left\langle \delta n^2\right\rangle l_p \sec\theta
\tag{4.57}
$$

$$
kl_p D_{\delta\overline{n}}(f(\Delta\rho)) = k^2\vartheta\left\langle \delta n^2\right\rangle l_p \sec\theta\left(1 - \varkappa\Re(f(\Delta\rho))\right).
\tag{4.58}
$$

With these substitutions and a change of variables, the following form is obtained:

$$
\begin{aligned}
\Gamma_{11}(x; \Delta\rho; \quad k_1, k_2) = {}& \exp\left\{-l_p\vartheta\left\langle \delta n^2\right\rangle l_p \sec\theta(k_1 - k_2)^2/2\right\}\cos\theta \\
& \times \iint \exp\left\{-l_p D_{\delta\overline{n}}(f'(\Delta\rho'))k_1 k_2\right\} \\
& \times \iint \exp\left\{-i\frac{\kappa'^2}{2\cos\theta}\left(\frac{1}{k_1} - \frac{1}{k_2}\right)\right\} \\
& \times \exp\left\{-i\kappa' \cdot (\Delta\rho' - \Delta\rho)\right\}\frac{d\kappa'}{(2\pi)^2}d\Delta\rho',
\end{aligned}
\tag{4.59}
$$

where the spatial correlation is defined by the modified form of $f(\Delta\rho)$ as distinguished by the prime

$$
f'(\Delta\rho) = \left(C\cos^2\theta\Delta\rho_y^2 - B\Delta\rho_y\Delta\rho_z + A\Delta\rho_z^2\right)/\left(AC - B^2/4\right).
\tag{4.60}
$$

The integral over κ' can be evaluated, which simplifies the result to

$$
\begin{aligned}
\Gamma_{11}(x; \Delta\rho; \quad k_1, k_2) &= \exp\left\{-l_p\vartheta\left\langle\delta n^2\right\rangle \sec\theta(k_1 - k_2)^2/2\right\} \\
&\times \iint \exp\left\{-l_p D_{\delta\bar{n}}\left(f'(\Delta\rho')\right)k_1 k_2\right\} \\
&\times \frac{\exp\left\{-\left[(\Delta\rho' - \Delta\rho)^T(\Delta\rho' - \Delta\rho)\right]/\alpha\right\}}{\pi\alpha} d\Delta\rho', \quad (4.61)
\end{aligned}
$$

where α

$$
\alpha = 2i\left(1/k_1 - 1/k_2\right)x \tag{4.62}
$$

captures the propagation distance dependence. The single-frequency result follows from the delta-function-like behavior of the Gaussian form as $\alpha \to 0$.

Because beacon satellite scintillation at frequencies below ~ 10 GHz is driven by ionospheric electron content structure, it is useful to recall the relation between refractive index and electron density derived in Section **??**, namely $\delta n = -2r_e\lambda\Delta N_e$. Thus, to accommodate plasma density fluctuations explicitly $k^2\left\langle\delta n^2\right\rangle$ is replaced by $4r_e^2\lambda^2\left\langle\delta N_e^2\right\rangle$. For scale-free power-law computations $k^2 C_s$ is replaced by $4\lambda^2\left(r_e^2 C_s\right)$ where C_s now applies to the SDF of the electron density structure rather than the refractive index structure. Additionally, for most applications, only single-point measurements ($\Delta\rho = 0$) are of interest. The result is also most informative when the frequency dependence of $\delta\bar{n}$ is written explicitly. To this end, the respective mean and difference frequencies are defined as

$$
\begin{aligned}
\bar{f} &= (f_1 + f_2)/2 &\quad (4.63) \\
\Delta f &= (f_1 - f_2)/2 &\quad (4.64)
\end{aligned}
$$

with $\bar{\lambda} = c/\bar{f}$, and $\bar{k} = 2\pi/\bar{\lambda}$. For practical applications $\bar{f}$ is the center or reference frequency and Δf is bounded by the system bandwidth. The relations will be discussed in more detail in Chapter 5. With these definitions and the translation from refractive index to electron density,

$$
\begin{aligned}
\Gamma_{11}(x; 0; k_1, k_2) &= \exp\left\{-4l_p r_e^2\vartheta\bar{\lambda}^2\left\langle\delta\bar{N}_e^2\right\rangle \sec\theta(\Delta f/\bar{f})^2/2\right\} \\
&\times \iint \exp\left\{-4l_p r_e^2\bar{\lambda}^2 D_{\delta\bar{N}_e}\left(f'(\Delta\rho')\right)\left(1 - \left(\Delta f/\bar{f}\right)^2\right)\right\} \\
&\times \frac{\exp\left\{-\left[\Delta\rho'^T\Delta\rho'\right]/\alpha\right\}}{\pi\alpha} d\Delta\rho', \quad (4.65)
\end{aligned}
$$

where

$$
\alpha = -i\left(\frac{\Delta f/\bar{f}}{1 - \Delta f^2/\bar{f}^2}\right)\frac{x}{\bar{k}}. \tag{4.66}
$$

Equation (12) in Rino et al. [45] neglected the displaced coordinate correction. The corrected definition replaces C with $C\cos^2\theta$ to conform to (4.60).

A number of techniques have been applied to produce practical analytic results. The early studies of the effects of frequency dispersion by Yeh and Liu, et al. [88], [89], [90], [61], relied on waveform moments to assess excess delay and pulse spreading. Approximations by Knepp, et al. [91] [92] allow direct use of a simplified frequency coherence function. However, as will be shown in Chapter 5, numerical integration is efficient and effective.

4.3.4 Spherical Wave Correction

The analytic calculations presented in Chapter 3 and the extension to oblique propagation paths in anisotropic media used plane-wave excitation exclusively. The plane-wave results capture scintillation effects as a modulation imparted to the propagation that evolves in the absence of structure. As such, they are not restrictive. However, for satellite observations path loss is critical in establishing signal strength and attests to the fact that wavefront curvature cannot be neglected. The incorporation of path loss is straightforward, but there is an additional scaling associated with spherical-wave propagation when structure is present.

In Appendix A.4, a simple but compelling development of the spherical-wave correction by Ratcliffe [72] is reproduced. The spatial variables (x and ρ or ς) that apply to the plane-wave results must be scaled by the spherical wave correction factor

$$SC = \frac{r_d}{r_d + r_p}, \tag{4.67}$$

where r_d is the propagation distance from the source to the disturbance and r_p is the distance from the disturbance to the receiver. Note that $SC <= 1$, and $\lim_{r_d \to \infty} SC = 1$, which is the plane-wave result.

The spherical-wave corrections affect plane-wave measurements in two ways. First, the replacement of the propagation distance $x \sec \theta$ by $r_p SC$ must be anticipated at the initiation of a computation. Second, the correction to the spatial scale can be accommodated by interpolating spatial fields generated by either theory or simulation.

The FPE formalism can accommodate real source excitation in lieu of the much simpler and less computationally demanding beam propagation. However, unlike the plane-wave ideal, the spherical-wave ideal is not so readily reproduced numerically. For example, a point source cannot be constrained to propagate only in the forward direction. It demands treatment in a spherical-wave system in which forward and backward spherical waves propagate inward and outward. Furthermore, reproducing source conditions for a satellite-borne transmitter requires information that is very difficult to obtain. Measurements of the antenna pattern on the spacecraft must be available and, to use them effectively, the attitude of the spacecraft at the point of transmission must be known. Fortunately, the spherical-wave correction is adequate for most applications.

4.4 SATELLITE ORBIT AND EARTH MAGNETIC FIELD CALCULATION

Theoretical results for satellite scintillation analysis are incomplete without utilities that can be used to compute the coordinate system origin, the source location relative to the receiver, and the anisotropy axis. These utilities come from decades of geolocation, satellite orbit measurement, and Earth magnetic field research. It is a propitious time to exploit these resources, which are now available on the Internet.

The constellation of Global Positioning System (GPS) satellites has revolutionized both time and position measurement. GPS latitude, longitude, and altitude measurements are often referenced to the World Geodetic System WGS84 ellipsoid, which is an agreed-upon surrogate for the Earth's surface. The geoid, which is the surface of zero gravitational potential, is also available in GPS coordinates. Moreover, very accurate measurements of the Earth's surface and identification layers are rapidly becoming available. GPS position requires very accurate timekeeping, which can be transferred to clocks on the devices that decode GPS signals. Part III of *Linear Algebra, Geodesy, and GPS* [93] provides a compact overview of GPS and coordinate systems, with a website that contains MATLAB utilities for geopositioning computations.

The Earth's magnetic field remains an active area of geophysical research. The Earth's main magnetic field evolves with time and is subject to transient perturbations that can disrupt electric power transmission and corrode large surface pipelines. Establishing the main magnetic field and disseminating it efficiently is an important activity. The International Geomagnetic Reference Field (IGRF) is a carefully managed spherical-harmonic representation of the Earth's main magnetic field used for this purpose.

4.4.1 Satellite Orbit Computation

The dissemination of satellite orbit prediction utilities predates GPS [94]. The community of U.S. Department of Defense and space research scientists needed an efficient means of predicting satellite orbits with sufficient accuracy to acquire satellite transmissions. This was achieved through the development of a Simplified General Perturbation (SGP) model, which is a system of modified Keplerian equations that can be solved efficiently using pseudo-orbital parameters calculated from high-precision orbital measurements. The necessary information is conveyed for SGP calculations via standardized sets of two-line elements. Each element set has a specified time epoch at which the orbital predictions apply most accurately. The results deteriorate with time, but generally provide sufficient accuracy for several days of high-resolution tracking. The results to be presented here were obtained with a slightly modified version of the SGP4 MATLAB code described in the Vallado, et al. report [94], which can be downloaded from the Internet.

The SGP4 code provides satellite position and velocity estimates versus time since epoch in an Earth-centered inertial coordinate system (ECI). Orbit calculations are performed in an inertial coordinate system that comprehends the relevant forces acting on the space object. Orbit calculation can be initiated with initial conditions derived from orbit measurements. However, the information an Earth-based observer requires is where and when to look for the satellite of interest. This requires both time and reference coordinate translation. Astronomers have confronted this problem for decades and adopted sidereal time, which is constructed to report the diurnal recurrence of an object fixed in space. It accounts for both the translational motion of the Earth over a one-day period and higher-order corrections as the applications demand. Consequently, sidereal time and geometric translations from the ecliptic plane must be used to convert the position and time in the ECI system to Earth-centered fixed coordinates (ECF) and universal time (UT); notationally $\mathbf{X}_{ECF}(UT)$ and $\mathbf{V}_{ECF}(UT)$.

Standard software utilities efficiently transfer $\mathbf{X}_{ECF}(UT)$ and $\mathbf{V}_{ECF}(UT)$ coordinates to GPS geodetic coordinates (latitude, longitude, and height with respect to the reference ellipsoid or to a local tangent-plane rectangular topocentric coordinate system (TCS). In the TCS system, positive x is eastward, positive y is northward, and positive z is normal to the reference ellipsoid at a specified origin. Rotations used in the standard utilities translate free vectors such as velocity among the coordinate systems. Upon converting the satellite ECF coordinates to a TCS system with origin at the point of reception, range, elevation, and true azimuth (bearing) to the satellite are readily computed as

$$r_{\text{sat}} \;=\; \sqrt{x_{TCS}^2 + y_{TCS}^2 + z_{TCS}^2} \tag{4.68}$$

$$\theta_{\text{sat}} \;=\; \arctan 2\left(z_{TCS}, \sqrt{x_{TCS}^2 + y_{TCS}^2} \right) \tag{4.69}$$

$$\phi_{\text{sat}} \;=\; \arctan 2\left(y_{TCS}, z_{TCS} \right) \tag{4.70}$$

At the surface of the reference ellipsoid the satellite is formally visible if θ_{sat} or z is greater than zero.[14]

Figure 4.3 shows an SGP4-derived C/NOFS satellite (NORAD Catalogue Number 32765) elevation-versus-time plot as observed from an equatorial station ($Lon = -44.2123°$, $Lat - 2.5934°$, $Alt = 48.20$ m). The triangles mark the start and end of the positive elevation segments, which are identified by arbitrary orbit numbers for reference. The beacon carried by the C/NOFS satellite was designed to measure ionospheric propagation disturbances. It broadcasts narrow-band signals at 150 MHz, 400 MHz, and 1067 MHz. A

---

[14]Satellite visibility at low or negative elevation angles is determined by local terrain conditions and the height of the antenna. Negative elevation is used as a cutoff for illustrative purposes.

high-elevation pass was chosen for illustrative purposes. Actual data from a similar equatorial pass will be discussed in Chapter 5.

Figure 4.4 shows the orbit geometry with respect to the station for Orbit 7, which reaches the highest elevation in the window shown in Figure 4.3. Velocity in the receiving station TCS coordinate system is obtained by rotating the velocity vector to the local cardinal directions (eastward, northward, and upward). With $\mathbf{r}_{TCS}$ and $\mathbf{v}_{TCS}$ representing the satellite range and velocity in the antenna TCS coordinate system, the line-of-sight direction to the satellite, $\mathbf{r}_{TCS}/r_{\text{sat}}$, and the range rate follow from the projection

$$\dot{r}_{\text{sat}} = (\mathbf{r}_{TCS}/r_{\text{sat}}) \cdot \mathbf{v}_{TCS}. \tag{4.71}$$

Doppler and range rate have opposite signs $(\dot{r} = -\lambda f_{Dop})$ in a reference coordinate system looking toward the object.

The disturbed region is generally confined to an altitude range that the satellite doesn't penetrate. For data analysis and modeling the line-of-sight intercept point at a specified altitude is used to identify the layer reference height. Computation of the intercept point searches using transformations from TCS to geodetic coordinates where altitude is measured directly. Figure 4.4 shows the geographic location of the 300-km intercept, which is not distinguishable from the orbit projection at the scale shown.[15] The velocity at the intercept point scales linearly with fractional path distance from the source to the satellite. Thus, the velocity of the origin of the reference coordinate system is readily computed.

4.4.2 Magnetic Field Computation

A subgroup of the International Union of Geodesy and Geophysics, namely the International Association of Geomagnetism and Aeronomy (IAGA), Division V-MOD, Geomagnetic Field Modeling, is charged with maintaining and disseminating a global magnetic field model. The current model release (IGRF10) can be downloaded from a NASA-sponsored website.[16]   At the time of writing this book the official model is available only in FORTRAN and C. The field directions are reported with x positive northward, y positive eastward, and z positive upward.

The model uses a spherical-harmonic expansion of the form

$$V(r, \phi, \theta) = a \sum_{l=1}^{L} \sum_{m=0}^{l} \left(\frac{a}{r}\right)^{l+1} (g_l^m \cos m\phi + h_l^m \sin m\phi) P_l^m (\cos \theta). \tag{4.72}$$

---

[15]The search is fast, but approximating the Earth surface segment with a sphere yields a faster analytic solution that is often used.
[16]The model generates the main field for dates between January 1, 1900 and January 1, 2010 in topocentric (or ECF) coordinates at geodetic locations for times in fractional years.

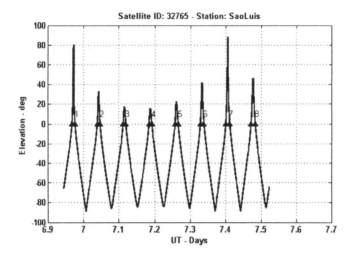

Figure 4.3 Elevation of C/NOFS satellite from an equatorial station. The triangles bracket positive elevation segments.

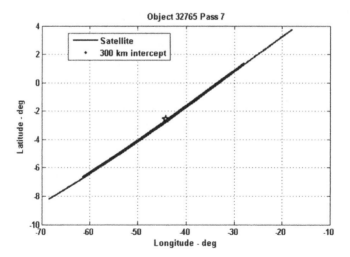

Figure 4.4 Satellite (solid) and 300-km intercept point (heavy dashed) trajectories with respect to receiving station (pentagram).

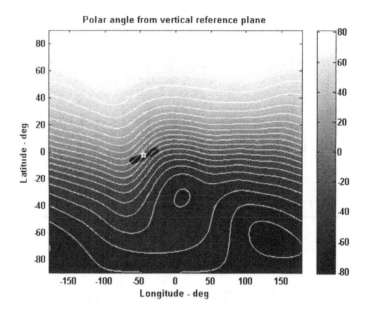

Figure 4.5   Global map of z component of magnetic field on the WGS84 reference ellipsoid.

The coefficients g_l^m and h_l^m that define the IGRF10 model are provided for each reference year in 5-year increments from 1900 to 2005. The years up to 2000 use 120 coefficients. The years between 2000 and 2005 have extended coefficient sets containing 195 coefficients. Linear interpolation is used to interpolate or extrapolate the field predictions. Figure 4.5 shows the z component of the IGRF10 main magnetic field at the surface of the Earth for the date 2000.2. The computation was performed with a MATLAB translation of the igrf10sys.for code obtained from the IAGA MOD-V website. The Sao Luis station is marked by the pentagram for reference. The penetration point geometry is marked with black crosses in the vicinity of $-5°$ Latitude, $-50°$ Longitude. The satellite location was chosen to probe the magnetic equatorial region.

4.5   EXAMPLES

Satellite scintillation structure as observed from near-Earth receiving stations depends on two sets of parameters. The first set is determined by the satellite orbit and the observing station location. It includes time-dependent values of r_{sat}, $\dot{r}_{sat}$, $\hat{a}_r = \mathbf{r}_{TCS}/r_{sat}$, the geodetic coordinates of the intercept point, $llh_{intercept}$, and the penetration point velocity, $\mathbf{v}_{TCS}$, for a specified reference coordinate system height. The calculation of the propagation system coordinates from the intercept point to the receiving antenna is done in two steps. First the coordinates of the station are calculated in a TCS coordinate centered on the intercept point. With these TCS coordinates, x_{TCS}, y_{TCS}, and z_{TCS}, the propagation system coordinates (x_P, y_P, z_P) are obtained by the reflection

$$
\begin{aligned}
x_P &= -z_{TCS} \\
y_P &= x_{TCS} \\
z_P &= -y_{TCS}
\end{aligned}
$$

depicted in Figure 4.1. The second set of parameters defines the statistical properties of the structure, which requires some elaboration. Once the structure is defined in the reference coordinate system, realizations are constructed and mapped onto the incremental path-integrated phase that drives the scintillation structure.

The anisotropy axis is aligned with the Earth's magnetic field at the origin of the reference coordinate system. Although the magnetic field elevation angle is nearly zero over the station, the magnetic field does admit a small variation during the west-to-east transit. This can be seen directly in Figure 4.6 where the variation of the Briggs-Parkins angle is shown. Perfect field alignment corresponds to a zero Briggs-Parkin angle or unity cosine. The change in aspect angle with respect to the magnetic field in the second half of the Orbit 7 pass corresponds to ~ 10 deg. As a general rule, an east west equatorial pass that cuts across the nearly horizontal magnetic field lines is considered to be ideal for measurement and interpretation of highly elongated equatorial irregularity structures.

With all the geometric factors established, structure characterization and realization can be addressed. Because the purpose here is to demonstrate methodology, no attempt is made to explore the very large parameter space that is of potential scientific and practical interest. Continuing with the scale-free power-law structure that is now very familiar, the large change of satellite position within the visible portion of the orbit requires large phase-screen realizations. The propagation distance dictates the maximum Fresnel radius that must be accommodated. Recall that fine sampling was used for the examples in Chapter 3 to accommodate extreme perturbation levels. However, the last example in Chapter 3 showed that the two-component power-law was dominated by the relatively benign shallow-slope regime. Consequently, the sample density can be decreased enough to accommodate the Fresnel radius at the largest ranges. Asymmetric grids that better match the structure anisotropy can also be used to advantage.

In Section 3.1 it was noted that models of neutral-atmosphere structure define the power-law index and turbulent strength or structure constant directly. Indeed, microwave radio devices called scintillometers are calibrated to measure the atmospheric structure constant directly. To frame ionospheric

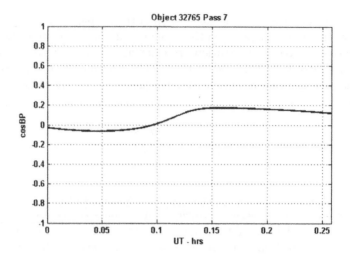

Figure 4.6   Briggs-Parkin angle variation with universal time for Orbit 7.

in terms of standard turbulence parameters, the ionosphere refractive index variance

$$\langle \delta n^2 \rangle = r_e^2 \lambda^4 \langle \delta N_e^2 \rangle / \pi^2 \tag{4.73}$$

is substituted into the defining relation

$$C_s = \frac{(4\pi)^{3/2} \Gamma(\nu + 1/2)}{\Gamma(\nu - 1) q_L^{-2\nu+2}} \langle \delta n^2 \rangle, \tag{4.74}$$

which yields the following measure of the turbulent structure supported by electron density variance up to the spatial wavenumber q_L:

$$C_s = r_e^2 \lambda^4 \frac{8\pi^{1/2} \Gamma(\nu + 1/2)}{\Gamma(\nu - 1)} q_L^{2\nu-2} \langle \delta N_e^2 \rangle. \tag{4.75}$$

The fact that the electron density structure varies with wavelength imposes the dramatic difference in wavelength sensitivity to atmospheric and ionospheric structure.

The electron density range for the ionosphere is roughly 10^9 to $10^{12}/m^3$. The upper frame of Figure 4.7 is a log-log plot of RMS electron density fluctuation and the C_s level at 400 MHz. For reference, the lower frame shows the corresponding SI level at 300 km. Normal incidence isotropic structure was used in the computation. The calculations show that a 1% fluctuation level in the mid-range of ionospheric electron density levels can saturate radio signals at UHF and lower frequencies.

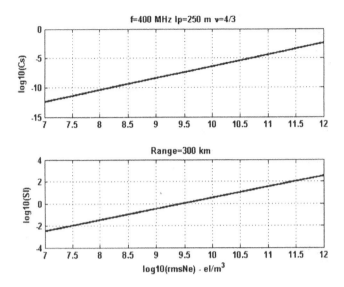

Figure 4.7   Log-log plots of turbulent strength (upper frame) and the log of SI (lower frame) versus RMS electron density.

4.5.1   Geometric Dependence of Anisotropy

The first two examples provide background material for interpreting the simulations results to be presented in Section 4.5.3. For the simulations, the geometry of Orbit 7 has been segmented into 47 paths separated by ~ 20 sec. The propagation along each path is simulated independently with phase-screen realizations that admit contours of constant correlation defined by (4.21). The contours are ellipses with axial ratios and orientations that vary in response to the changing propagation geometry. Figure 4.8 shows the anisotropy variation for 10:1 elongation. For display clarity, only the anisotropy ellipse for every other path is displayed. Each ellipse is plotted at the origin of the rectangular propagation coordinate system. The initial intercept is marked with a pentagram. Since the first satellite-to-receiver path is west of the receiving station (See Figure 4.4), the measurement-plane intercept in the propagation coordinate system is east of the propagation system origin at the center of the grid. The satellite pass cuts across the overhead field lines at a near normal angle ($\psi_{BP} \sim \pi/2$). The slow variation of the orientation of the ellipses from path to path reflects the change in the azimuth angle of the magnetic field.

4.5.2   Geometric Dependence of Intensity Scintillation

The scintillation index has already been introduced as a defining parameter for scintillation structure. Large structure development differences between the

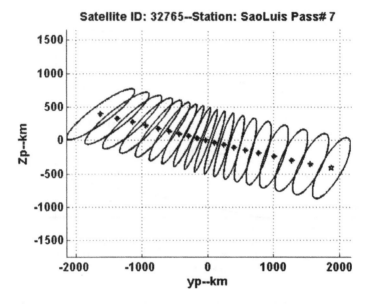

Figure 4.8   Measurement plane anisotropy plotted in the propagation coordinate system with positive y eastward and positive z southward from the layer penetration point.

small-slope and large-slope regimes within the index parameter range $0.5 < \nu < 2.5$ have been demonstrated.   Oblique incidence and anisotropy do not change this general characterization.   However, as the medium becomes two-dimensional, the partitioning from small-slope to large-slope behavior shifts to more shallow slopes.   For now, consider the weak-scatter analytic result

$$SI^2 = \sec\theta C_p \rho_F^{(2\nu-1)} \frac{\Gamma\left((2.5-\nu)/2\right)}{2^{(\nu+1/2)}\sqrt{\pi}\Gamma\left((\nu+0.5)/2\right)(\nu-0.5)}\varpi, \qquad (4.76)$$

where

$$\rho_F = \sqrt{SCr_p/k}. \qquad (4.77)$$

Note that the definition of the Fresnel scale, ρ_F, includes the correction for the finite source distance with SC defined by (4.67).

Figure 4.9 shows model computations of the SI variation at 150 MHz for the Orbit 7 equatorial satellite pass.   The computation was performed for isotropic and 10:1 anisotropic structures.   The geometric variation for the elongated irregularities is nearly two-dimensional, which would be strictly true if the magnetic field orientation with respect to the propagation direction did not vary.   The fact that the same turbulent strength produces weaker intensity scintillation with elongation is a consequence of the normalization that preserves refractive index variance for the same C_s level.   Values of

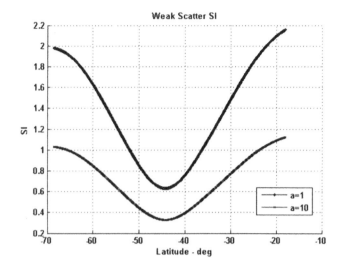

Figure 4.9 Weak scatter prediction of geometric variation of SI for Orbit 7 pass with isotropic structure and 10:1 elongated irregularities.

SI approaching or exceeding unity are outside the range of the weak-scatter theory but are indicative of strong-scatter behavior.

Although the results here are for a single frequency, the weak-scatter theory in the scale-free limit predicts the frequency dependence $SI \sim f^{-(\nu+1.5)/2}$. Experience has shown that this frequency dependence significantly underpredicts the observed equatorial frequency dependence. For example, the equatorial scintillation at GHz frequencies observed by Craft and Westerlund [95] was unexpected. It was later shown that multifrequency observations could be reconciled by invoking a two-component power-law SDF as discussed in Section 3.6.3 of Chapter 3 and the references cited therein.

4.5.3 Beacon Satellite Simulations

This section builds on the geometric relations developed to accommodate anisotropy and oblique propagation to simulate the measurement-plane fields representative of equatorial beacon satellite scintillation. The computational requirements for a single transition of a disturbed medium are not increased significantly from the normal-incidence isotropic structure that was simulated in Chapter 3. However, the potential parameter space of interest has increased significantly. Fortunately, the scale-free power-law range collapses the critical dependencies into a few critical parameters. Indeed, in Chapter 5 it is shown that two-dimensional simulations capture the essential structure for system performance analysis.

4.5.3.1 Background The FPE split-step recursion for 3-D isotropic structures at normal incidence is defined by (3.123), (3.124), and (3.125). The same recursion framework is used for propagation through anisotropic structure at oblique incidence with the following adjustments. In the propagator, the x component of the spatial wavenumber is computed for each value of κ with the displaced transverse wave vector and the κ-dependent offset

$$kg(\kappa) \to kg(\kappa + \mathbf{k}_T) - (\tan\theta)\,\widehat{\mathbf{a}}_{k_T}\cdot\kappa. \tag{4.78}$$

Realizations of the phase perturbation are calculated with the replacements

$$C_s \quad \to \quad C_s ab\sec\theta \tag{4.79}$$

$$\Phi_{\delta:}(\kappa) \quad \to \quad C_s \left(A\kappa_y^2 + B\kappa_y\kappa_z + C\kappa_z^2\right)^{-(2\nu+1)}. \tag{4.80}$$

The notation δ: means δn (atmosphere) or δN_e (ionosphere) to emphasize that the simulation capability is general even though only the ionosphere is used here. Indeed, the phase structure can be added to a uniform or varying background. These changes are straightforward and impose no significant increase in the computational requirements. The challenge is to provide a viable definition of the propagation space.

In the simulation examples presented in Chapter 3, the two-dimensional complex field realization is available for analysis. Thus it was possible to construct radial wavenumber spectral estimates or measure the complex field directly. The amplitude and phase of the complex field are defined by the equivalence

$$\psi(y,z) = A(y,z)\exp\{i\phi(y,z)\}. \tag{4.81}$$

If the complex field were the actual starting point, the interpretation would be complicated by the fact that the representation is unchanged if any multiple of 2π is added to $\phi(y,z)$. A continuous phase map, akin to the phase structure that initiates the simulation, can be extracted by first evaluating

$$\arctan 2(\mathrm{Im}(\psi(y,z)), \mathrm{Re}(\psi(y,z))),$$

which constrains the result to lie between $-\pi$ and π radians. A continuous record is reconstructed by imposing a constraint on the change that can occur between adjacent data samples. For two-dimensional complex fields, algorithmic procedures have been developed to perform this operation [96].

Single-point measurements, however, are recorded as one-dimensional time series. Phase reconstruction is straightforward, but necessary only as a means of simulating the ideal output of a frequency/phase tracking receiver. Phase jumps occur where the signal amplitude approaches zero. These details are best studied with simulations, but it is useful to review the connection between the FPE theory and more familiar scintillation developments. Under the narrow-angle scatter approximation, the spectrum of waves that propagate beyond a phase screen at $x = 0$ is

$$\widehat{\psi}_{\mathbf{k}}(x;\boldsymbol{\kappa}) = \widehat{\psi}_{k}(0;\boldsymbol{\kappa})\exp\left\{i\left(\kappa^{2}+\tan^{2}\theta\left(\widehat{\mathbf{a}}_{k_{T}}\cdot\boldsymbol{\kappa}\right)^{2}\right)\frac{x\sec\theta}{2k}\right\}, \qquad (4.82)$$

where

$$\begin{aligned}\widehat{\psi}_{k}(x_{0};\boldsymbol{\kappa}) &= \iint \exp\{i\delta\phi\left(\boldsymbol{\rho}'\right)\}\exp\left\{-i\boldsymbol{\kappa}\cdot\boldsymbol{\rho}'\right\}d\boldsymbol{\rho}'\\ &\simeq 2\pi\delta\left(\boldsymbol{\kappa}\right)+i\widehat{\delta\phi}\left(\boldsymbol{\kappa}\right),\end{aligned} \qquad (4.83)$$

$$A_{\psi_{k}} \simeq 1 + 2\,\mathrm{Im}\left(\widehat{\delta\phi}\left(\boldsymbol{\kappa}\right)\right), \qquad (4.84)$$

and

$$\phi_{\psi_{k}} \simeq \mathrm{Re}\left(\widehat{\delta\phi}\left(\boldsymbol{\kappa}\right)\right). \qquad (4.85)$$

The intensity SDF is the same as (4.19) less the singularity at the origin. The corresponding phase SDF is

$$\Phi_{\phi}(\boldsymbol{\kappa}) = \Phi_{\delta\phi}(\boldsymbol{\kappa})\cos^{2}\left(\left(\kappa^{2}+\tan^{2}\theta\left(\widehat{\mathbf{a}}_{k_{T}}\cdot\boldsymbol{\kappa}\right)^{2}\right)\frac{x\sec\theta}{2k}\right). \qquad (4.86)$$

As already demonstrated, the intensity SDF is strongly suppressed at low spatial wavenumbers, while the phase spectrum envelope is unchanged. The Rytov approximation gives the same result for the phase SDF. However, the standard development suggests that the range of validity is not rigorously constrained to small phase excursions [9, Chapter 6.5].

For completeness, a standard SDF estimate derived from uniformly spaced data samples, v_{k}, can be constructed as

$$\widehat{\Phi}_{n} = \frac{1}{N_{s}}\sum_{m=0}^{N_{s}-1}\left|\frac{1}{N_{\mathrm{fft}}}\sum_{k=0}^{N_{\mathrm{fft}}-1}w_{k}v_{k+(m-m_{0})N_{s}}\exp\left\{-2\pi ink/N_{\mathrm{fft}}\right\}\right|^{2}, \qquad (4.87)$$

where w_{k} is a symmetric weighting function normalized such that

$$\sum_{k=0}^{N-1}|w_{k}|^{2} = 1. \qquad (4.88)$$

The number of segments to be averaged (with overlap if $m_{0} > 0$) is N_{s}. If N_{fft} exceeds the number of data samples in the segment, the data are zero padded to N_{fft}. If N is the number of data samples, one can show that

$$\sum_{k=0}^{N-1}\left\langle|w_{k}v_{k}|^{2}\right\rangle = \sum_{n=0}^{N_{\mathrm{fft}}-1}\left\langle\widehat{\Phi}_{n}\right\rangle. \qquad (4.89)$$

To the extent that the average spectral structure changes slowly compared to the resolution of the spectrum of the window function w_k,

$$\left\langle |v|^2 \right\rangle \simeq \sum_{n=0}^{N_{fft}-1} \left\langle \widehat{\Phi}_n/N \right\rangle. \tag{4.90}$$

It follows that $\widehat{\Phi}_n/N$ is the spectral intensity per SDF sample. To scale the measured SDF for comparison to theoretical computations, $\widehat{\Phi}_n/N$ should be divided by the spatial or temporal frequency interval.

4.5.3.2 Measurement-Plane Realizations   The simulations use 2048-by-8192 point single phase-screen realizations sampled at 10 wavelengths per sample. The change in sample density and asymmetric sampling was used to better match satellite geometries and the structure asymmetry. Aside from the change in sampling and the incorporation of anisotropy and oblique propagation, the split-step realization of the complex field in the measurement plane is the same as the normal-incidence isotropic simulations presented in Chapter 3. Figure 4.10 shows the simulated measurement-plane intensity in dB units for the first of the 47 paths that comprise Orbit 7. The diagonal striation of the bands of intensity enhancement and the periodic continuity of the structure at the edges is exploited to avoid edge effects. The simulation conserves the initial average plane-wave signal intensity, which is normalized to 1 (zero dB). The orientation of the striations is consistent with the rightmost ellipse shown in Figure 4.8. Another feature to note is the tendency for strong scintillation to occur in alternating bands even though the source structure is nominally homogeneous.

Under the frozen-field assumption, the measurement plane translates with velocity $\mathbf{v}_k$. Thus, a point measurement can be simulated by interpolating the complex field to the spatial points defined by $\boldsymbol{\rho}_n = \mathbf{v}_k(n\Delta t)$. The transit is from the lower right-hand corner of the field realization to the upper left-hand corner along the trajectory highlighted in Figure 4.10. The simulated intensity record is shown in Figure 4.11, where the upper frame shows the intensity in dB versus time in seconds from the start of the scan, and the lower frame shows SI over 0.75-sec intervals using 512 points. The systematic change in the structure level from moderate to weak to strong quantifies what is observed in the intensity realization. Although these simulations look very much like real data, theory predicts a uniform SI level corresponding to the fixed turbulence parameters that initiated the structure.

The simulated phase is summarized in Figure 4.12. The upper frame shows the reconstructed phase obtained by removing the arc-tangent 2π ambiguities. The mean phase computed over the same interval used to compute SI is overlaid in the upper frame. The phase residual associated with random scintillation structure is shown in the lower frame. The phase data confirm that the regions of enhanced SI correspond to local large-scale phase enhancements.

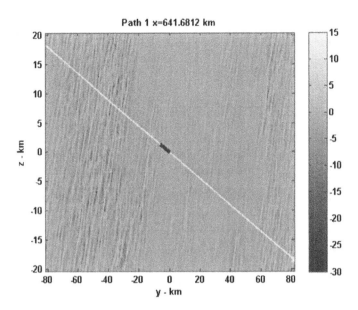

Figure 4.10 Measurement-plane intensity variation for Path 1 of Orbit 7. The diagonal line is the scan direction through the origin.

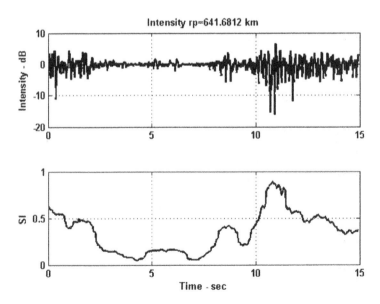

Figure 4.11 Path 1 raw scintillation record (upper frame) and sliding 512-sample estimate of SI (lower frame).

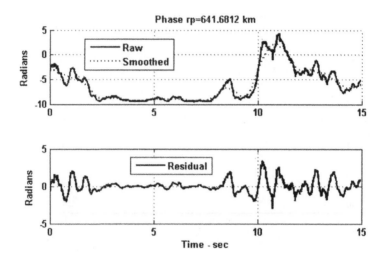

Figure 4.12 Reconstructed phase (upper frame dashed) from scan of Path 1 measurement-plane field. The solid line in the upper frame is a running average with its residual shown in the lower frame.

Figures 4.13, 4.14, and 4.15 show, respectively, the measurement-plane intensity, the measured SI variation, and the reconstructed phase variation for Path 23 of Orbit 7. The most conspicuous difference between Path 1 and Path 23 is the finer field sampling due to the much slower effective scan rate. The overall scintillation levels are lower as well, reflecting both the shorter path through the structure and the shorter propagation distance to the receiver. The alternating bands of enhanced intensity and phase fluctuation persist. The reconstructed phase shows generally the same overall characteristics as the Path 1 phase.

An independent phase screen realization was used for each path simulation. Random seeds were used because the intercept point moves a significant but varying transverse distance as the satellite traverses its orbit. The 60-sec duration implied by the slow scan velocity, however, is not realizable with a set of uniform point measurements over the entire pass. As already noted, a more faithful mapping of structure together with a more dense path sampling would allow interpolation through changing structures, but no new insight would be gained.

4.5.3.3 Spectral Analysis The spectral index and the turbulent strength have been emphasized as robust parameters to characterize power-law processes. However, conventional spectral analysis emphasizes the trade between resolution and uncertainty. It is well known, for example, that accurate estimation

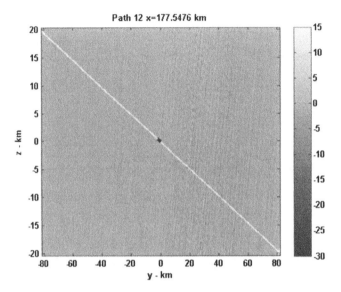

Figure 4.13   Measurement-plane intensity variation for Path 23 of Orbit 7.   The diagonal line is the scan direction through the origin.

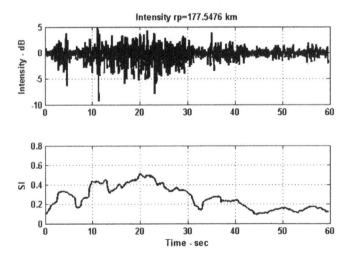

Figure 4.14   Path 23 raw scintillation record (upper frame) and sliding 512-sample estimate of SI (lower frame).

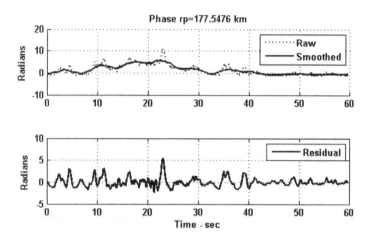

Figure 4.15   Reconstructed phase (upper frame dashed) from scan of Path 23 measurement-plane field. The solid line in the upper frame is a running average with its residual shown in the lower frame.

of uniform spectral intensity level requires segmentation and averaging, which amounts to trading resolution for improved uncertainty.   Since the realizations that produced the intensity records in Figures 4.11 and 4.14 were generated with uniform power-law structures, a segmentation and averaging scheme was anticipated.   However, any meaningful segmentation of the data record yields dissimilar data segments that can't be averaged to improve statistical measures.   This is a defect of correlation measures that was highlighted in Chapter 3.

Clearly, spatial-domain measures are difficult to interpret quantitatively. Spectral-domain measures, by comparison, intrinsically separate the structure by spatial scale, which is the inverse of spatial frequency. This suggests that the SDF might provide a more direct measure of the driving point structure conditions.   This is supported by Figure 4.16, which shows intensity and phase spectra computed over the entire 8192-sample interval.   The overlay designated theory was computed from (4.54) with the nominal parameters that were used to initiate the phase screen realization.   For example, v_{eff} is computed from (4.48) using the known geometric and model parameters used to generate the phase screen realizations.   The low-frequency suppression of the intensity spectrum expected from Fresnel filtering is evident in the intensity spectrum.   The low-frequency distortion of the phase spectrum is attributed to the use of a Hamming window to suppress the effects of edge discontinuities, which are pronounced in the raw phase data.

The qualitative agreement with theory does not address estimation of the critical power-law parameters T and ν.   It is well-known that increasing the

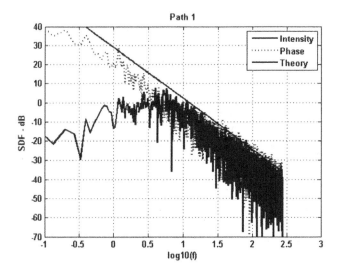

Figure 4.16 Intensity and phase PSD estimates derived from the entire Path 1 data interval.

duration of the measurement interval does not decrease the uncertainty in estimating a uniform SDF level, which is the basis for estimates of the SDF using (4.87). It is also known that power-law processes behave differently ([97] and [98]). Wavelet scale spectra have been demonstrated to give both improved estimates of the underlying power-law parameters and direct measures of the locally homogeneous range. Wavelets effectively provide a scale-dependent frequency decomposition. The wavelet scale spectrum, which is an average over the scale-dependent frequency estimates proposed by Hudgins [99], achieves the improved statistical stability. However, a demonstration of the forward consistency of the simulations with the underlying theoretical spectra is sufficient.

To summarize, a persistent feature of intensity scintillation in power-law environments that has been attributed to self-scaling is manifest as local intensity scintillation enhancements. Spectral domain measures, particularly wavelet-based measures, should provide quantitative links to the underlying structure.

4.6 THEORY AND SIMULATIONS

The material in this chapter completes the development of the theory of scintillation, which is embodied in a first-order differential equation, the FPE, that can be integrated to characterize the cumulative propagation effects up to the current computation plane. The weakly inhomogeneous media structures that the FPE accommodates include homogeneous media and smoothly varying media with or without the small-scale structure usually associated with scintillation. Power-law media, with moderate constraints on the range of the power-law index, admit a scale-free regime for which measures of turbulent strength and the spectral index characterize the potential severity in any cumulative propagation disturbance. If the structure is concentrated in a thin layer, solutions for the principal moments that characterize the field structure admit simple algebraic representations. Because of this ease of computation, phase-screen results provide important analysis guidelines. This was emphasized by the selection of simulation examples in Chapters 3 and 4.

The simulations provided two important insights. First, although the phase screens that were used to impose the structure on the wave fields were derived from homogeneous ensembles, the intensity scintillation showed a non-uniform structure. This was attributed to a scale-sampling effect in which focusing from large-scale but random lens-like structures create local regions of enhanced scintillation. These local departures from homogeneity are more prominent in highly anisotropic media where the enhancements appeared in bands, but they are also apparent in scintillation from isotropic media. The second important result demonstrated that the spectral distribution of the structure, as defined by the power-law envelope predicted by the path-integrated phase structure, can be recovered from SDF measures over intervals that would not be deemed homogeneous in data-driven segmentation.

Equally important, simulations can provide insight into the effects of varying background structure, and as well as varying structure distributions along the principal propagation path. The simulations provide numerical data for which the driving point conditions are known precisely but admit no viable analytic characterization. This situation is common in computational electromagnetics. Electromagnetic field realizations are computed for ideal scattering objects that are characterized by boundary conditions on a specified surface. The computation methods are often validated by comparing the predictions to exacting laboratory measurements. Disparities are typically attributed to differences between the real and the ideal model, not computational errors. The integrity of the computation is demonstrated by consistency checks.

Propagation environments are more demanding because they cannot be measured completely with the level of detail that simulations can accommodate. Whereas point-by-point validations of measured and simulated fields are not feasible, simulations do provide a means to validate robust measures of in situ structure details related to the physical processes that were hypothesized to explain the development of structure at representative disturbance levels.

Most published simulations have been contrived to reproduce theoretical predictions. However, insofar as the FPE is concerned there is no reason to constrain structures that physical processes might dictate. The last example in Section 2.2.2 showed that propagation along a highly elongated structure need not be excluded. One practical concern has been finessed by allowing edge structure to repeat periodically. However, to the extent that significant energy is carried by waves whose central ray intercepts the computation boundary, the energy reentering the computation grid can alter answers from those that used a larger computation grid. In effect, one has a proper FPE solution that is not physical. There are techniques that can be applied to remove energy flowing into a computation boundary. The simplest is a gradual field taper. More accurate schemes are computationally demanding. For unbounded scintillation environments this is usually not a concern. Physical boundaries will be discussed in Chapter 6.

The primary challenge in attacking large-scale problems is developing a complete characterization of the environment. It is rarely possible to reconstruct the dynamic evolution over an extended propagation region. Nonetheless, considerable progress has been made in isolating and modeling the dominant physical processes. Modeling the nonlinear evolution of the nighttime equatorial structure known as plumes or bubbles depends on how they are observed. It is possible to simulate the structure development over very large physical volumes with very high resolution. Even if this were feasible to the scale range that causes scintillation, managing the data volume would be very challenging. With some effort, however, the simulation methodology developed in Chapters 2, 3, and 4, could be applied to this problem.

CHAPTER 5

SYSTEM APPLICATIONS OF SCINTILLATION THEORY

Information is the resolution of uncertainty.

—*Claude Shannon*

The development of scintillation theory in Chapters 2, 3, and 4 characterized the propagation of time-harmonic fields in weakly inhomogeneous media. Slow time variations induced by the combined motion of sources, receivers, and media structure were incorporated as translations of the spatial structure. However, systems that exploit wave propagation to transmit and extract information necessarily use sequences of waveforms that typically occupy much broader bandwidths. This was anticipated with the inclusion of frequency coherence among the statistical measures that were introduced in Chapters 3 and 4. This chapter completes the development with the introduction of a framework that can accommodate the full range of propagation disturbances that affect communication and surveillance systems. The development starts with a review of waveform structures, signal processing, and fundamental limitations imposed by additive noise. Signal processors also may have to contend with interference, jamming, and clutter, but the concern here is the impact of propagation disturbances that impart a random frequency-dependent complex modulation onto the signal that effectively competes with the modulation introduced for remote sensing or information transmission.

In information theory, a channel is the physical connection between an information source and its destination.   Information theory also tells us that the accuracy of information transfer depends only on the signal bandwidth and the signal-to-noise ratio.   Channels that exploit wave propagation invariably use frequency translation to transmit the information as efficiently as possible.   The details of the frequency translation and the modulation and demodulation operations used to extract information-bearing signals are reversible and need not be considered explicitly.   The linearity of the propagation equations facilitates modeling the channel as a linear combination of the Fourier components that comprise the information-bearing waveforms.   Thus, a channel model provides an analytic framework to analyze all information transmission and remote sensing systems.   In remote sensing the information is imparted onto the transmitted waveform by media interactions that occur during transmission.   A channel model also provides a framework to design signal processing algorithms for scintillation measurement.

Because of the complexity and interplay of signal processing and detection operations, numerical simulations invariably provide the most effective means to evaluate system-specific performance or to execute a design-phase trade study.   It is important to note that, aside from absorption, forward propagation effects are formally irreversible.   Thus, there is a possibility to reconstruct the undisturbed wave field and use that information to mitigate deleterious effects.   The reconstruction of blurred optical images, synthetic aperture radar images, and channel equalization are examples of this type of mitigation.   In comparison, a deep fade can momentarily disrupt the operation of a system. Operational diversity is the most effective mitigator of this type of disturbance, but diversity is expensive to implement, whereby a modified system design may be necessary to alleviate the effects of propagation disturbances. (See Table 5.1 for symbols.)

5.1   AN INTRODUCTION TO WAVEFORMS

The FPE formalism allows us to characterize a complex wave field that has propagated in a weakly inhomogeneous medium from a point of transmission to a measurement plane. To be specific, consider the complex wave field $\psi_\mathbf{k}(\boldsymbol{\rho} + \mathbf{v_k}t; \mathbf{R}_0, f_c + f)$ where $\psi_\mathbf{k}(\boldsymbol{\rho}; \mathbf{R}_0, f)$ is a solution to the FPE from a source propagating along the reference direction defined by $\mathbf{k}$. The phase center of the receiving antenna is at position $\boldsymbol{\rho}_n$ in a reference plane centered at $\mathbf{R}_0$. The apparent drift velocity of the stationary field structure is $\mathbf{v_k}$, and $f_c + f$ is the frequency relative to the reference transmission frequency f_c.

To study system effects it will be necessary to characterize the modulating waveforms, $\widehat{v}_m(f) \Leftrightarrow v(\tau)$, where the symbol $\Leftrightarrow$ denotes a baseband frequency transformation. Baseband means about zero as opposed to the actual frequency content about the carrier frequency f_c. These concepts will be developed in detail shortly. Looking ahead, the linearity of the FPE ensures that a modulation at the source creates a signal with spectral content $\widehat{v}_m(f)\psi_\mathbf{k}(\boldsymbol{\rho} + \mathbf{v_k}t; \mathbf{R}_0, f_c + f)$. Depending on how $\psi_\mathbf{k}(\boldsymbol{\rho}; \mathbf{R}_0, f)$ is computed, it may or may not contain path loss over the propagation distance from the source to $\mathbf{R}_0$. However, plane wave solutions provide a means to isolate the deterministic channel elements that can be evaluated analytically. Since channel modeling applications address the propagation-induced modulation, it is assumed here that $\psi_\mathbf{k}$ represents the disturbance imparted to wave field. Corrections for path loss and wavefront curvature can be incorporated as described in Chapter 4.

Channel models were introduced initially to characterize multipath as encountered in radar echoes from large randomly located scattering centers [100]. The tie to scintillation is the conceptual picture of multiple ray paths connecting the source and receiver. Whereas scattering channels are inherently

Table 5.1   Chapter 5 Symbols

Symbol	Definition
$p(\tau) \Leftrightarrow \widehat{p}(f)$	Baseband waveform/Fourier spectrum
$v_m(t)$	Transmitted waveform (5.1)
$\widehat{v}_m(f, d)$	Fourier transform of $v_m(t)$ (5.9)
$D(f, d)$	Frequency-dependent Doppler shift (5.10)
$h(t; f_c + f)$	Channel transfer function (5.12)
$s_m^n(\tau)$	Channel output with noise (5.15)
$P_\mathrm{D}, P_\mathrm{FA}$	Single waveform detection/false alarm probability

dispersive because of the relatively large delay spread, scintillation is predominantly a temporal modulation. In effect, the ray paths cannot be resolved. Channel models have a long history of use to characterize the system effects of propagation disturbances. The applications include the effects of signal absorption, delay errors, Doppler errors, and scattering losses. Straightforward channel model enhancements can accommodate environmental noise, interference, discrete multipath, and clutter. The source of multipath and clutter is the transmitted signal itself. The theoretical framework to incorporate multipath and clutter will be presented in Chapter 6. In this chapter, only one-way propagation disturbances will be considered explicitly.

5.1.1   Signal Structure

Signals designed for information transmission and remote sensing can be represented as complex waveform sequences. Invoking the principle of time retardation, a transmitted waveform sequence upon reception admits a mathematical representation of the form

$$v_m(t) = p_m(t - r(t)/c - mT) \exp\{2\pi i f_c(t - r(t)/c)\}, \tag{5.1}$$

where t represents time, $r(t)/c$ represents the instantaneous propagation delay to a point in the propagation space at range $r(t)$, and m identifies a specific waveform in the transmission sequence. The waveform repetition interval, T, is fixed, but the complex waveform $p_m(t)$ can be a sequence of sub-pulses that use phase, frequency, or pulse position to convey information. The frequency spectrum of the waveform $p_m(t)$ is confined to BW about the center or carrier frequency f_c. The narrowband representation (5.1) requires the frequency interval $f_c - BW/2 < f < f_c + BW/2$ to be nonnegative and generally small compared to f_c.[17]

To accommodate continuous modulation representative of broadcast AM and FM signals, a model takes the form

$$v(t) = M(t - r(t)/c) \exp\{2\pi i f_c(t - r(t)/c)\}, \tag{5.2}$$

where $M(t)$ is a continuous complex modulation imparted to the carrier signal. It will be shown that this is exactly the form of non-dispersive scintillation. The annoying effects of signal fading and distortion are examples of random modulation imparted to an information-bearing waveform. Digital signal processing offers considerable opportunity to compensate the effects of channel-induced distortions.

---

[17]The apparent inconsistency between time-finite and band-limited signals is resolved by recognizing that the time-bandwidth product identifies the number of significant orthogonal minimum-duration minimum-bandwidth signals that represent the signal [101].

5.1.2   Signal Processing

In the real world, one cannot simply apply a matched filter to maximize the signal-to-noise ratio at the matched filter output because the propagation delay, $\tau = r(t)/c$, and the frequency, f_c, are critical unknown parameters. They are needed for signal alignment and phase reference, respectively. Range can be estimated by multiplying the measured delay by an average propagation velocity. If the positions of the source and the receiver are known, ray tracing using a parameterized refractive-index model can determine the range that connects the point of transmission and the point of reception.[18] Alternatively, the departure of the ray path from the reference direction can be used to reverse the process and estimate refractive conditions that best explain the data.

If the significant spectral content of the waveform occupies the frequency band $f_c - BW/2 < f < f_c + BW/2$, sampling at twice the Nyquist frequency, $BW/2$, is sufficient to recover the waveform. Thus, an upper bound on the finest time scale for signal processing is

$$\delta t = 1/BW. \tag{5.3}$$

Oversampling has advantages that sometimes merit the increased sampling burden. In any case, the next critical time interval is the waveform duration $T_p \leq T$. The length of time that the waveform is close to its peak intensity, $T_{p'}$, establishes the duty cycle measured by the fraction $T_{p'}/T$. The duty cycle is important because it establishes the minimum average power the transmitter must deliver. If the waveform is captured in a background of uncorrelated (white) noise, the time-bandwidth product $T_{p'} \times BW$ defines the number of independent noise samples that contribute to each waveform sample. The maximum signal-to-noise ratio at the output of an appropriately constructed matched filter is increased potentially by the time-bandwidth product. This favors the use of sub-pulses with long duration; however, $1/T$ (the Nyquist rate) must be large enough to support recovery of the signal Doppler content.

Range variation with time due to changing geometry was introduced in Chapter 3. Over the transmission time of N waveforms, the range change admits a Taylor-series representation of the form

$$r(t) = r_0 + \dot{r}(t - t_0) + \ddot{r}(t - t_0)^2/2 + \dddot{r}(t - t_0)^3/6 + \cdots . \tag{5.4}$$

Substituting (5.4) truncated at $\dot{r}$ into (5.1), it follows that

$$\begin{aligned} v_m(t_0 + t) &= p(t_0 + t - (r_0 + \dot{r}t)/c - mT) \\ &\quad \times \exp\{-2\pi i f_c(t_0 + t - (r_0 + \dot{r}t)/c)\}. \end{aligned} \tag{5.5}$$

The consequence of significant higher-order terms will be discussed shortly. However, there is no loss of generality in taking $t_0 = r_0/c$ with $t = \tau + mT$,

---

[18]See (2.27).

whereby

$$v_m(t_0 + t) = p(\tau(1 - \dot{r}/c) - \dot{r}mT/c)\exp\{-2\pi i f_c\tau(1 - \dot{r}/c)\}$$
$$\times \exp\{2\pi im(f_cT/c)\dot{r}\}\exp\{-2\pi imf_cT\}. \tag{5.6}$$

With $\dot{r}/c \ll 1$, and an integral number of f_c cycles in the interval T, the signal model simplifies to

$$v_m(\tau) = p(\tau - \dot{r}mT/c)\exp\{-2\pi if_c\tau\}$$
$$\times \exp\{2\pi im(f_cT/c)\dot{r}\}. \tag{5.7}$$

An integral number of f_c samples in the interval T effectively means that the time base for transmission and reception are identical. Many signal processing presentations start with the assumption that (5.7) is valid, but this is achieved only with good engineering.

The signal model represented by (5.7) establishes the minimum interval, T, that supports linear range-rate change. There are situations where higher-order terms must be taken into account, but only the linear model will be considered here. Most signal capture operations assume that the pulse-to-pulse change in range is well-represented by a Doppler shift over all N waveform transmissions. The essential point here is to establish a fast-time variable, τ, that incorporates the fine detail of the waveform, including the frequency offset $\exp\{-2\pi if_c\tau\}$. In a radar system the pulse delay, r_0/c, and range rate, $\dot{r}$, are parameters to be estimated. In a communication system, delay and Doppler tracking are necessary for signal decoding. Either way, the linear model is more complicated than one might expect from the relatively simple form of (5.7).

To demonstrate the interplay between delay and Doppler, consider the frequency content of $v_m(\tau)$ as defined by the Fourier transformation

$$\hat{v}_m(f, d) = \int v_m(\tau)\exp\{-2\pi if\tau\}d\tau. \tag{5.8}$$

This is the modulation that would be extracted by an ideal receiver with phase-locked sampling. It follows by direct computation using (5.7) that

$$\hat{v}_m(f, d) = \int_0^{T_p} p(\tau - \dot{r}mT/c)\exp\{-2\pi i(f - f_c\dot{r}/c)\tau\}d\tau$$
$$\times \exp\{2\pi if_CmT\dot{r}/c\}$$
$$= \int_{-\dot{r}mT/c}^{T_p-\dot{r}mT/c} p(u)\exp\{-2\pi i(f - f_c\dot{r}/c)u\}d\tau$$
$$\times \exp\{-2\pi iD(f, d)mT\}$$
$$= \hat{p}(f + d)\exp\{-2\pi iD(f, d)mT\}, \tag{5.9}$$

where $d = -f_c\dot{r}/c$ is the true Doppler shift and

$$D(f, d) = (1 - f/f_c)d + d^2/f_c \tag{5.10}$$

Table 5.2   Signal Processing Time Scales

Symbol	Definition	Comments
Δt	Delay sample interval	$\Delta t = 1/BW$
τ	Pulse duration	$\geq \Delta t$
T	Pulse repetition interval	Doppler Nyquist $(1/(2T))$
CPI	Coherent processing interval	$M \times T$
T_{Dwell}	Dwell time	$N \times CPI$
NCI	Non-coherent integration	Dwell level operation

is the Doppler shift that would be extracted from a DFT computation applied to a set of complex signal samples separated by the pulse repetition interval T. An astute reader will note that there are two true Doppler frequencies that map to the same apparent Doppler shift. This would occur only if the Doppler shift were a substantial fraction of the carrier frequency. As a practical matter, the linear term in (5.10) defines the coupling between frequency and Doppler, which is the frequency-domain manifestation of the time-domain coupling between delay and Doppler referred to as range walk in radar systems. In effect, the time of each waveform reception creeps a bit depending on the range rate.

For most applications the Doppler shifts and fractional bandwidths are small enough that $D(f, d)$ can be replaced by d or at most $(1 - f/f_c)d$. Furthermore, the Doppler shift of the waveform usually can be ignored in the design of the matched filter. For most signal processing operations

$$\widehat{v}_m(f) \simeq \widehat{p}(f) \exp\{-2\pi i dmT\} \tag{5.11}$$

implies that a single matched filter equal to $\widehat{p}^*(f)$ can be employed for waveform delay measurement or synchronization. Pursuing this detailed interplay between delay and Doppler might seem pedantic, but modern digital signal processing capabilities present practical situations where the fine details of the signal structure and departures from the linear model become important. SAR processing, which will be discussed in Chapter 6, often requires range-walk correction.

In summary, signal capture for measurement or information extraction purposes involves a hierarchy of time scales. The longer time scales become important when received waveforms undergo processing operations that combine partially processed subsets. Table 5.2 contains a reference list of signal processing time intervals.

5.2 SCINTILLATION CHANNEL MODEL

The channel model for propagation disturbances characterizes baseband waveforms referenced to the measurement plane. As already noted, the linearity of the FPE implies that the summation of frequency-dependent solutions that comprise an inverse Fourier transformation is also an FPE solution. Thus, the scintillation *channel transfer function* can be defined as

$$h(t; f_c + f) = \left(\sqrt{CF}/R_0\right) \psi_{\mathbf{k}}(\boldsymbol{\rho} + \mathbf{v_k}t; \mathbf{R}_0, f_c + f), \qquad (5.12)$$

where the factor $\left(\sqrt{CF}/R_0\right)$ has been introduced to convert the amplitude to square root SNR units as described in Section 1.2. It is understood that $\psi_{\mathbf{k}}(\boldsymbol{\rho}; \mathbf{R}_0, f)$ is a solution to the FPE with plane wave excitation and spherical wave corrections applied. The channel transfer function notation $h(t; f_c + f)$ suppresses all but the variables that a signal processor comprehends explicitly. As already noted, to analyze angle errors in a large continuous or synthetic aperture system it may be desirable to include the spatial variable, $\boldsymbol{\rho}$, explicitly as well. The path loss variation over the range of $\boldsymbol{\rho}_n$ is generally negligible. For discrete array processing, particularly for adaptive processing, it is important to note further that the Doppler shift can vary from element to element across the array.

Whatever the particular application, by using (5.11) the received signal from the mth waveform transmission admits the representation

$$
\begin{aligned}
\overline{v}_m(\tau, mT) &= \int h(mT, f_c + f)\widehat{p}(f) \exp\{2\pi i f \tau\} df \\
&\quad \times \exp\{-2\pi i dmT\}.
\end{aligned}
\qquad (5.13)
$$

One can use (5.9) to estimate the amount of frequency decorrelation across the band that can be tolerated for a particular waveform. If the frequency content of the transmitted signal is much narrower than the frequency range over which scintillation structure changes, which is usually the case, (5.13) can be represented as a complex modulation applied to the waveform:

$$
\begin{aligned}
\overline{v}_m(\tau, mT) &\simeq h(mT, f_c) \int \widehat{v}_m(f) \exp\{2\pi i f \tau\} df \\
&= h(mT, f_c) v_m(\tau).
\end{aligned}
\qquad (5.14)
$$

This is usually referred to as frequency-flat fading because the frequency dependence of the scintillation is negligible over the waveform frequency band.

The random structure of the complex modulation can still change both the amplitude and phase of the signal on the time scale at which pulse-to-pulse Doppler estimates are made. However, because there is no distortion of the waveform itself, $v_m(\tau)$ can be replaced by the pulse-compressed waveform at the output of the matched filter without loss of generality. To complete the general channel model, additive noise is included. The simplest form of a complete channel model is

$$s_m^n(\tau) = \bar{v}_m(\tau) + \varsigma^n(\tau), \tag{5.15}$$

where $\varsigma_n(\tau)$ represents uncorrelated unit variance background noise, and $\bar{v}_m(\tau)$ is represented by (5.14). If the channel frequency coherence is larger than the waveform bandwidth,

$$s_m^n(\tau) \simeq h_n(mT, f_c)v_m(\tau) + \varsigma_n(\tau + mT). \tag{5.16}$$

If the additive noise is not uniform across the frequency band, a whitening operation can be applied prior to any subsequent signal processing. For most applications, an equivalent signal representation of the form (5.16) establishes a performance baseline.

5.2.1   Applications to Non-Dispersive Fading

To illustrate the use of the channel model for performance evaluation, a simple application that is amenable to complete analytic computation will be reviewed. Assume that the channel supports the full bandwidth of the signal, whereby (5.16) applies. Assume further that each waveform has been sampled at the peak output of the matched filter, formally

$$P(mT) = \max \left| \int \hat{s}_m^n(f) \hat{v}_m^*(f) \exp\{2\pi i f\tau\} dt \right|^2. \tag{5.17}$$

From (5.16), the operation is statistically equivalent to taking the magnitude of the sequence of complex numbers

$$s_m = h_m v_m + \varsigma_m, \tag{5.18}$$

where the information content resides in v_m, and h_m represents the channel-induced modulation sampled at intervals T, and ς_m is unit-variance white Gaussian noise.

The interpretation of this result is important. The implicit assumption is that the complex signal is subject only to the uncertainty introduced by ς_m. Signal strength has already been enhanced by the time-bandwidth product through the coherent matched filter operation (5.17). To isolate the signal in frequency and further enhance the detectability of $h_m v_m$, additional coherent processing can be used. For example, if h_m is constant over N waveform samples, the Fourier transformation of the sequence has the form

$$\hat{s}_m \simeq h_m \hat{v}_m + \hat{\varsigma}_m. \tag{5.19}$$

Because the Fourier transformation is linear, $\hat{\varsigma}_m$ is still uncorrelated, but the relative strength of the signal has increased by approximately the number of samples in the FFT. Typically, small coherent processing losses are incurred because of the finite processing interval and signal tapering imposed to minimize Doppler sidelobes.

How scintillation affects coherent processing is evaluated most effectively by generating realizations of h_m. In doing so, however, it is necessary to impose both the first-order statistics and the appropriate waveform-to-waveform channel coherence to the realizations. Although the analytic results derived in Chapter 3 can be used to estimate the coherence times, imposing an appropriate fade distribution is more difficult. The ideal approach would be to generate realizations of the channel transfer function. To reduce the computation load, one-dimensional phase screens are most often used for such applications ([92] and [102]). Knepp's approach preserves the essential characteristics of the first- and second-order statistics. The two-dimensional simulations allow a large number of realizations to be averaged, thereby reducing statistical uncertainty in the performance measures. The importance of simulations is underscored by the fact that performance depends on the processing and detection strategies that are used.

Early radar studies of the effects of pulse-to-pulse time variation by Peter Swerling have become analytic standards to evaluate the effects of random waveform modulations on the probability of waveform detection. A summary of the Swerling model types and their extensions can be found in *Fundamentals of Radar Signal Processing* [103, Section 6.3.4]. All the Swerling model computations use stair-step waveforms to approximate the random modulation by constant levels over the processing intervals summarized in Table 5.2. For example, if the amplitude change over a dwell is negligible, Doppler estimates can be made that are not affected by scintillation.

5.2.1.1 Non-Fading Channels A sufficient statistic for non-coherent detection is the summation

$$y = \sum_{m=1}^{M} |\widehat{s}_m|^2 / \sigma_N^2, \tag{5.20}$$

where $\sigma_N^2 = \left\langle |\widehat{s}_m|^2 \right\rangle$ is the noise variance. The normalization imparts SNR_c intensity units to y. The c subscript indicates that the post processing SNR is to be used. It is assumed that prior to formation of the detection statistic, the signal has been match filtered and Fourier transformed, which potentially achieves the full coherent processing gain implied by the time-bandwidth product times the number of pulses used in the Doppler extraction. If there is no scintillation, the probability density function (PDF) of the test statistic y is the Rice distribution

$$p(y|X) = \left(\frac{y}{X}\right)^{(M-1)/2} \exp\{-(y+X)\} I_{M-1}\left(2\sqrt{yX}\right), \tag{5.21}$$

where $X = SNR_c$. It can be shown that

$$\langle y \rangle = (M + X) \tag{5.22}$$

$$\langle y^2 \rangle - \langle y \rangle^2 = 2(M + X). \tag{5.23}$$

These results can be obtained directly from the signal model or by integrating the PDF multiplied by the appropriate power of y. The $X = 0$ case has already been encountered in Chapter 3 as a form of the intensity distribution. Here it represents the null-hypothesis probability (no signal present):

$$\lim_{X \to 0} p(y|X) = y^{(M-1)} \exp\{-y\}/\Gamma(M). \tag{5.24}$$

Assuming that the phase is randomly distributed even after coherent processing has been applied may be confusing. The lack of coherence applies strictly with respect to the noise. Suppose, for example, the noise contribution in phase with the signal could be isolated. Then the quadrature noise could be eliminated and the detection statistic would be a purely real Gaussian variable with positive and negative excursions about its mean. Fully coherent detection of this type is used in high performance communication systems and cell phones. Even so, a more conservative non-coherent noise model is used most often.

Standard detection probability calculations for non-fading detections in a background of Gaussian noise are obtained from the cumulative distribution

$$P_D(M, X, t) = \int_t^\infty \left(\frac{y}{X}\right)^{(M-1)/2} \exp\{-(y + X)\} I_{M-1}\left(2\sqrt{yX}\right) dy, \tag{5.25}$$

where t is the detection threshold. This is the probability that the test statistic will exceed the threshold when the signal is present. The false alarm probability is

$$P_{\text{FA}}(M, t) = \int_t^\infty y^{(M-1)} \exp\{-y\}/\Gamma(M). \tag{5.26}$$

The threshold can be chosen to achieve a desired false alarm rate. Figure 5.1 shows the non-fading P_D for a quadratic detector in a white noise background. The threshold was set to achieve a false alarm rate of 10^{-8}. The detection parameters were chosen to reproduce a result analyzed in [103, Figure 6.10].

5.2.1.2 Slowly Fading Channels Under fading conditions, the following average detection probability is used:

$$\overline{P}_D = \int_0^\infty P_D(N, x, T) p(x|\overline{X}, p1, ...) dx, \tag{5.27}$$

where $p(x|\overline{X}, p1, ...)$ is the SNR distribution with $\overline{X}$ representing the mean SNR, which is now a conditioning variable, and $p1, ...$ represents additional parameters such as the scintillation index. To evaluate (5.27) the parameterized

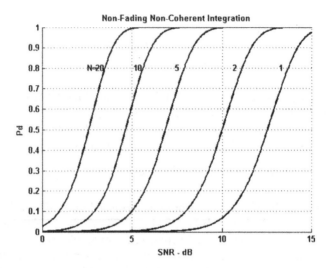

Figure 5.1   Probability of detection versus post-processing SNR after non-coherent integration of M processed samples.

intensity PDF is needed. As discussed in Chapter 3, the scintillation intensity PDF is not a direct product of the theory. No single model adequately covers the full range of fading conditions. The Nakagami distribution, which is the simplest distribution that has been used, provides fading characteristics that are distinctly different from the log-normal distribution, which generally fits measured weak-to-moderate scintillation very well. In the saturation regime, $SI > 1$, the K-distribution is more accurate.

Figures 5.2 and 5.3 show the evaluation of (5.27) for the Nakagami distribution and the log-normal distribution, respectively, as they are defined in Section 3.5.1. Below $SI = 1$, the Nakagami distribution has a higher probability of fades below the mean intensity than the corresponding log-normal distribution. Consequently, the Nakagami $SI = 1$ fading loss is comparable to the loss for log-normal fading at $SI = 2$. At low SNR levels fading actually improves performance, while the gain at high signal levels is larger than the loss during fades. As the fading becomes more severe, the fade duration is not accurately reproduced by the stair-step approximation that underlies (5.27). It is then more accurate to use simulations based on time-series representations of the fading signal structure.

5.3   SYSTEM PERFORMANCE ANALYSIS

There are numerous models that predict the performance of communication and remote sensing systems with varying degrees of fidelity. The applications

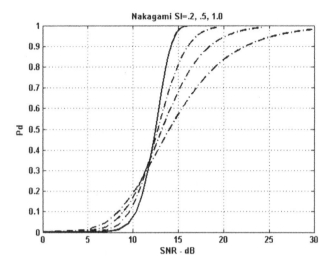

Figure 5.2   Probability of detection for Nakagami fading distribution with $SI = .2$, .5, and 1.

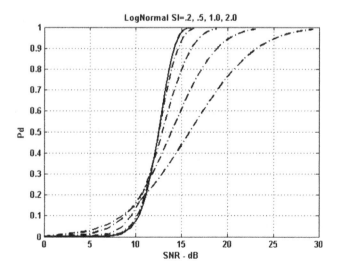

Figure 5.3   Probability of detection for log-normal fading distribution with $SI = .2$, .5, 1, and 2.

include satellite communication, surveillance, and remote sensing. The frequency range covers VHF to visible light. The models accommodate a broad range of atmospheric meteorological disturbances, which begin to dominate at frequencies above 1 GHz. However, the frequency dependence of the refractive index in the ionosphere induces excess delay and, to a lesser extent, Doppler variations that must be accommodated. Beyond that, ionospheric scintillation is a concern in systems that use long temporal processing intervals. The objective here is to identify analysis procedures that can be used as building blocks to achieve the level of fidelity required by specific applications.

5.3.1   System Sensitivity and Processing Intervals

How a system will perform in a given situation is driven primarily by SNR. It will be shown shortly, for example, that scintillation parameter estimation requires a processed SNR approaching 20 dB. Practical considerations invariably use both hardware elements and signal processing to achieve the necessary SNR. SNR computation starts with (1.40) and (1.41). In the design phase, trade studies usually determine a suitable balance between cost and performance. It is assumed here that the waveforms and the bandwidth have been determined. Preprocessing refers to waveform operations that result in a data sequence from which the desired information can be extracted. Where necessary, preprocessing includes calibration and equalization. Specific system requirements can be realized with additional preprocessing gains. Preprocessing operations also include frequency tracking and time alignment.

A well-designed system invariably has a known noise-limited performance baseline against which the achieved performance can be compared. This is particularly true of communication systems where error rates are readily measured and monitored. The baseline performance of a surveillance system is often specified as the detection probability versus range for a compact target of known cross section. Calibrated targets are routinely used to monitor performance. Communication and surveillance systems also use multiple modes of operation. For example, a radar may give up resolution for sensitivity at longer ranges. It may also operate with coarse search modes that can be switched to high resolution modes for classification. Propagation effects are often discovered when expected performance is degraded or unexpected signals appear.

Whatever the specific configuration and application might be, performance analysis starts with knowledge of SNR and preprocessing gains that exploit processing over the time intervals listed in Section 5.1.2. The deleterious effects of propagation disturbances then follow as the noise-limited gains are compromised by channel-induced degradation.

5.3.2 Coherence Bandwidth

Matched filtering presupposes that the transmitted waveform is known in detail. However, even under ideal conditions, transmission and detection operations introduce frequency distortion. Modulating a high-power transmitter invariably introduces nonlinearities that distort the waveform. Multielement broadband antennas introduce dispersion because the active phase center varies with frequency. Filters that are used to confine transmitted signals, isolate received signals, and limit noise can also introduce frequency distortions. The systematic dispersion induced by the frequency dependence of the ionospheric refractive index also introduces excess delay, Doppler, and in some cases waveform distortion. Good engineering practice can compensate for these measurable distortions, but channel-induced random frequency distortions remain.

The two-frequency mutual coherence function described in Chapter 3 and its extension in Chapter 4 were introduced as a measure of random frequency distortion. The leading multiplicative factor in (4.65) is a residue of the frequency dependence of the refractive index. It is important because it affects the scale-free limit. However, in the spectral index range $0.5 < \nu < 1.5$, the integral term dominates and the following integral coherence measure derived from (4.65) can be used:

$$
\overline{\Gamma}_{11}(x; 0; \Delta f/\overline{f}) = \iint \exp\left\{-C_{\delta\phi}\vartheta \sec\theta f(\Delta\zeta)^{2\nu-1}\left(1 - \Delta f/\overline{f}\right)\alpha^{\nu-1/2}\right\}
$$

$$
\times \frac{\exp\left\{-\Delta\zeta^{T}\Delta\zeta\right\}}{\pi}d\Delta\zeta. \tag{5.28}
$$

The defining parameters are repeated here for reference:

$$
\overline{f} = (f_1 + f_2)/2 \tag{5.29}
$$

$$
\Delta f = (f_1 - f_2)/2 \tag{5.30}
$$

$$
\overline{\lambda} = c/\overline{f}, \overline{k} = 2\pi/\overline{\lambda} \tag{5.31}
$$

$$
\alpha = -i\left(\frac{\Delta f/\overline{f}}{1 - \Delta f^2/\overline{f}^2}\right)\frac{x}{\overline{k}}. \tag{5.32}
$$

The frequency dependence enters in three terms, namely the fractional frequency separation $\Delta f/\overline{f}$, the perturbation strength dependence on the mean wavelength, and the propagation term α. The parameter $C_{\delta\phi}$ is the phase structure constant and C_p is the phase turbulent strength, which have been defined previously as

$$
C_{\delta\phi} = \frac{C_p\Gamma(3/2 - \nu)}{\pi\Gamma(\nu + 1/2)(2\nu - 1)2^{2\nu-1}} \tag{5.33}
$$

$$
C_p = \overline{\lambda}^2 r_e^2 l_p C_s \tag{5.34}
$$

A variable change was applied to (4.65) to collect all critical parameters in a single argument. As $\Delta f/\overline{f} \to 0$, the leading exponential approaches unity, and the remaining Gaussian term integrates to unity. That is,

$$\lim_{\Delta f/\overline{f} \to 0} \overline{\Gamma}_{11}(x; 0; \Delta f/\overline{f}) = 1. \tag{5.35}$$

To sustain a reduction in coherence bandwidth, it is necessary that

$$\text{Re}\left\{ \sec\theta C_{\delta\phi}\vartheta \left(1 - \Delta f/\overline{f}\right) \alpha^{\nu - 1/2} \right\} > 1. \tag{5.36}$$

The following coherence bandwidth measure can be used for the shallow-slope scale-free regime:

$$\Delta f/\overline{f} \simeq \frac{\cos\left(\frac{\pi}{2}\left(\nu - 1/2\right)\right)^{\frac{1}{\nu - 1/2}}}{\left(\sec\theta C_{\delta\phi}\vartheta\right)^{\frac{1}{\nu - 1/2}} \rho_F}. \tag{5.37}$$

For a fixed power-law index, reducing the fractional bandwidth requires increasing the denominator, which is proportional to $SI^{2\nu - 1}$. The geometric picture is that signals propagating over two paths separated enough to produce a deep fade will cause the replicas of the waveforms propagating along these paths to interfere destructively.

Figure 5.4 shows a numerical integration of (5.28) at 400 MHz at 10 km from the phase screen. Theoretical results based on (5.4) have been shown to be in agreement with multi-frequency equatorial scintillation beacon data ([45]). As such, the results can be used to identify conditions under which coherence bandwidth loss might result in significant performance degradation. When the coherence bandwidth loss is significant, numerical simulations can be used to assess the actual performance loss ([104] and [92]).

5.3.3 Temporal Coherence

Temporal coherence loss degrades coherent processing over intervals that use multiple waveforms. The degradation of the single-pulse probability of detection due to fading was demonstrated in Section 5.2.1. Spreading information over multiple waveforms that fade independently is the classic way to mitigate fading losses, but slow fading is very difficult to overcome because of the large blocks of waveforms that must be processed. Fortunately, modern satellite communication systems operate at frequencies where scintillation effects are generally benign, although excess delay and Doppler corrections remain important. The GPS satellite system, which operates at L-band, experienced ionosphere-induced fading at equatorial tracking stations that was severe enough to cause the receivers to lose code synchronization. Receiver modifications mitigated this problem, but scintillation can still limit GPS performance under extreme ionospheric disturbance conditions.

Loss of temporal coherence remains as the primary concern for most operational systems. Satellite systems at L-band and lower frequencies that

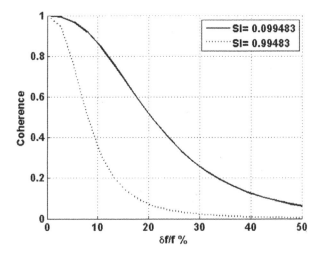

Figure 5.4   Two-frequency coherence as a function of fractional frequency separation for weak (solid) and strong (dotted) intensity scintillation.

operate with long coherent processing intervals are most vulnerable. These include VHF and UHF airborne communications and satellite synthetic aperture radar operations, which will be discussed after the introduction of two-way propagation in Chapter 6.

The remainder of this section will address temporal coherence loss. A measure of temporal coherence similar to (5.36) can be derived from (4.49) as

$$|\Delta t| \simeq \frac{1}{\left(C_{\delta\phi} v_{eff} \vartheta \sec\theta\right)^{\frac{1}{2\nu-1}}}, \tag{5.38}$$

but the ramifications of temporal coherence loss in systems that use coherent processing is difficult to ascertain without simulations. The simulations require a representative waveform sequence and a large number of independent realizations of the channel-induced modulation represented by (5.14). To minimize the computational load it may be desirable to use the two-dimensional form of the FPE introduced in Chapter 2. Some motivation is provided by the equatorial scintillation structure shown in Chapter 4, which is essentially two-dimensional. To extract the relevant the two-dimensional geometry, x_P, y_P, z_P must be rotated to a magnetic-field-aligned system. Using the same procedure as presented in Section 4.2.1, let

$$\begin{bmatrix} \overline{x}_P \\ \overline{y}_P \\ \overline{z}_P \end{bmatrix} = U_\psi U_{\phi_B} \begin{bmatrix} x_P \\ y_P \\ z_P \end{bmatrix}. \tag{5.39}$$

In the translated displaced coordinate system,

$$R_{\delta\phi_k}(\Delta\overline{\rho}) \simeq k^2 l_p \sec^2\overline{\theta} \int_{-\infty}^{\infty} R_{\delta n}\left(\Delta\eta, \Delta\overline{\rho} - \tan\overline{\theta}\widehat{\mathbf{a}}_{\overline{k}_T}\Delta\eta\right) d\Delta\eta. \tag{5.40}$$

If one assumes there is no variation along the $\overline{z}_P$ axis, the spatial Fourier transformation of $R_{\delta\phi_k}(\Delta\overline{\rho})$ reduces to

$$\begin{aligned}
\Phi_{\delta\phi_k}(\kappa_y) &= k^2 l_p \sec^2\overline{\theta}\Phi_{\delta n}\left(\tan\overline{\theta}\cos\overline{\phi}\kappa_y, \kappa_y\right) \\
&\simeq k^2 l_p \sec^2\overline{\theta}C_s\left(1 + \tan^2\overline{\theta}\cos^2\overline{\phi}\right)^{-(2\nu+1)}\kappa_y^{-(2\nu+1)}. \tag{5.41}
\end{aligned}$$

Note also that when $\overline{\phi} = 0$, the κ_y dependence scales with $\sec^2\overline{\theta}$, which effectively eliminates the angle dependence, as desired for propagation in the magnetic field plane.

It is important to keep in mind that there is no strict two-dimensional equivalence. One must accept the fact that when the underlying physics supports two-dimensional structure evolution rather than the three-dimensional structure evolution assumed in Chapters 3 and 4, the results are distinctly different. For example, a measurement of the one-dimensional spectrum derived from a PSD estimate would reflect the in situ power-law index directly, rather than the one-dimensional phase index, $p = 2\nu$, derived in Chapter 4. As a rule, results derived in Chapter 4 that remain finite as the axial ratio $a \to \infty$ can be applied directly. Those that do not remain finite must be rederived for the strictly two-dimensional model. As an alternative, the FPE could be applied directly and without geometric constraint to highly elongated structures. This would be necessary, for example, to model propagation at high geomagnetic latitudes where nearly field-aligned propagation can occur.

For the two-dimensional problem, because the geometric factors in (5.41) enter only as turbulent strength scale factors, there is no loss of generality in considering only normal incidence. Thus, the one-dimensional model introduced in Chapter 2 can be used directly, but the small-slope to large-slope behavior demonstrated in Chapter 4 occurs over different spectral index ranges (see also [105]). Thus, two examples are presented, one with $\nu = 2$ to enhance the large scale structure, and the second with $\nu = 1$ to be more noise like. Consistent with the channel model, noise is added to the simulated wave field. The noise level was set to generate an SNR of 30 dB in the spectral domain. Figures 5.5 and 5.6 show propagation structure at ~ 316 km from realizations of $\nu = 2$ and $\nu = 1$ power-law structures, respectively. The upper frames show the signal intensities in dB. The lower frames show the complex data samples plotted against their real and imaginary components.

The simulated data are presented in the same format as the scintillation example in Chapter 2. The spatial extent of the data is 5 km. For typical satellite trajectories and irregularity drifts, coherent processing intervals might comprise 512 to 1024 samples. The parameters for this example were chosen to admit channel-induced fluctuations within a 512-sample coherent

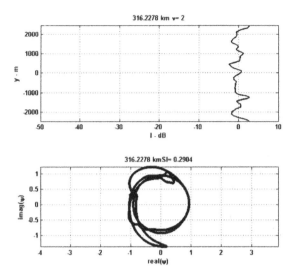

Figure 5.5   Measurement-plane intensity and phase for one-dimensional phase screen with $\nu = 2$.

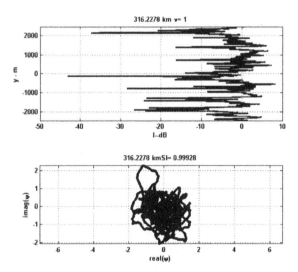

Figure 5.6   Measurement-plane intensity and phase for one-dimensional phase screen with $\nu = 1$.

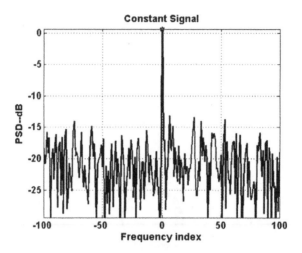

Figure 5.7   Power spectral density of 512-sample segment with constant signal.

processing interval. To provide a baseline reference for noise-limited coherent processing gain, Figure 5.7 shows the PSD of a constant signal (no scintillation) at the mean intensity level embedded in noise. The PSD was scaled to place the processed coherent signal peak at 0 dB. The PSD was also computed with zero-padding to 8-times the number of samples in the coherent processing interval. This was done to emphasize that frequency-measurement accuracy is limited by the noise, not the sample rate. This will be discussed in more detail in the following section.

Replacing the constant signal that produced the reference PSD shown in Figure 5.7 with the one dimensional complex wave field simulates the distortion that would result if the matched filter outputs from each waveform were being used for coherent processing. The upper frames of Figures 5.8 and 5.9 show the results of this replacement, again using the first 512 samples of realizations of the phase screens that produced the measurement-plane fields in Figures 5.5 and 5.6, respectively. The lower frames show the associated phase scintillation. The examples have very different intensity scintillation levels, as can be seen by comparing the upper frames in Figures 5.5 and 5.6. The steeper index ($\nu = 2$) produces a moderate coherence loss in spite of the large phase excursions, whereas the large slope ($\nu = 1.0$) essentially negates the coherence gain. At perturbation levels approaching saturation, the coherence is completely lost and cannot be recovered with increased SNR.

The examples presented here illustrate the unmitigated losses that scintillation can introduce. Although the examples represent the degradation of preprocessing a single waveform, a 10 dB loss of coherent processing gain is significant. However, it has been noted that mitigation procedures can recover

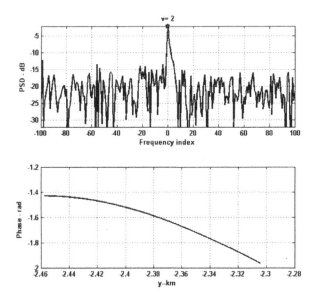

Figure 5.8 PSD of first 512-sample CPI (upper frame) and signal phase (lower frame) for $\nu = 2$ data shown in Figure 5.5.

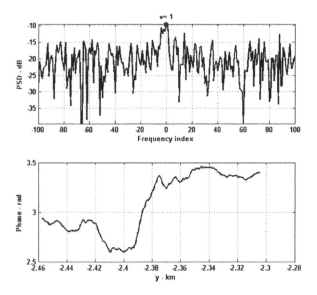

Figure 5.9 PSD of first 512-sample CPI (upper frame) and signal phase (lower frame) for $\nu = 1.0$ data shown in Figure 5.6.

losses that are phase dominated by recognizing that each phase measurement contains information and propagation-induced structure. Depending on the nature of the information content, phase errors can be adaptively compensated.

5.3.4 Spatial Coherence

Spatial coherence is a straightforward generalization of temporal coherence. Indeed, one proceeds conceptually from the two-dimensional measurement plane structure to a time series that represents a single receiver channel measurement. The structure of the two-dimensional field was developed and demonstrated in Chapter 4. However, it should be noted that spatial coherence and angular characterization of scattering are intimately related. The two-dimensional spatial wavenumber SDF can be converted to angle with the transformation

$$\phi_p = \tan^{-1}(\kappa_y, \kappa_z) \tag{5.42}$$

$$\theta_p = \cos^{-1}\left(\sqrt{1 - (\kappa/k)^2}\right). \tag{5.43}$$

To the extent that there is significant scatter outside the beam of the receiving antenna, that energy is lost and is usually reported as a *scattering loss* [106].

Explicit reference to angular scattering is more commonly used in optical systems where two-dimensional measurements are the norm. Angular characterization is also used in radio astronomy where angle of arrival also is a more fundamental measurement. The essential point is that the FPE formulation supports whatever field characterization is appropriate to the particular measurements. The simulations in Chapters 4 and 5 particularly for an optical system operating in turbulent media and extraterrestrial radio astronomy provide the structural details for analysis and exploitation of spatial coherence.

5.4 SCINTILLATION DATA PROCESSING

Processing satellite transmissions to extract amplitude and phase scintillation is intimately related to assessing system performance. The main difference is that waveforms designed for channel measurement are usually narrowband with low data rate modulation that is easily separated from the scintillation modulation. The critical dependence on SNR will be demonstrated directly. However, the emphasis here is on scintillation measurement as a critical first step in scintillation analysis. Once a reliable measure of the complex modulation is obtained, model-dependent interpretation of the parameters follows the analysis presented in Chapter 4.

5.4.1 Background

A narrowband signal is ideally suited for scintillation measurement. Because of the narrow bandwidth of the signal, frequency dispersion is not a concern. Indeed, exploring frequency coherence requires a set of narrowband signals

[6]. The signal model (5.16) can be rewritten as

$$v(t) = h\left(t; f_c\right) s(t) + n\left(t\right).$$ (5.44)

The narrowband signal itself is well approximated by

$$s(t) = A \exp\left\{2\pi i f_c\left(t - r(t)/c\right)\right\},$$ (5.45)

where A represents a constant amplitude. The term $h\left(t; f_c\right)$ represents the complex modulation imparted to the signal during transmission, propagation, and reception. Transmission and reception are emphasized here because they affect the interpretation of propagation effects. The additive term $n\left(t\right)$ represents the background noise, whose main contribution usually comes from the receiver low-noise amplifiers. It is assumed that the receiver limits the bandwidth of $v(t)$ to accommodate both the Doppler spread of $h\left(t; f_c\right)$ and the maximum expected Doppler shift. Aliased signals can be identified and corrected, but for now assume the signals of interest lie within the receiver passband.

Consider the Fourier transform of $v(t)$ over the contiguous coherent processing intervals $t_0 + nT < t < t_0 + (n+1)T$. If $h\left(t; f_c\right)$ is nearly constant over these intervals,

$$\widehat{v}_T(f) \simeq h\left(t_0 + nT; f_c\right) \widehat{s}_T\left(f\right) + \widehat{n}_T\left(f\right).$$ (5.46)

To the extent that $r(t_0 + t) = r\left(t_0\right) + \dot{r}\left(t_0\right)\left(t - t_0\right)$, the signal spectrum can be approximated as

$$\widehat{s}_T\left(f\right) \propto A\delta\left(f - f_c\dot{r}\left(t_0\right)/c\right).$$ (5.47)

Assume further that an estimate of the average noise passband is available and has been used to whiten (5.46), whereby $\left\langle\left|\widehat{n}_T\left(f\right)\right|^2\right\rangle = 1$, and $\widehat{v}_T(f)$ has square root of SNR units in the frequency domain, hereafter denoted $\sqrt{SNR}$. A signal processing algorithm that captures $\widehat{v}_T(f)$ at $f = f_D = -f_c\dot{r}\left(t_0\right)/c$ presents output consisting of a signal plus additive white noise. The intensity of the processed signal peak in (5.46) stands above the noise by the coherent processing gain, which approaches $10\log 10\left(T/\Delta t\right)$.

Before describing the signal capture algorithm, however, it is instructive to assess the effect of additive white noise on estimating the scintillation index. Consider the following scintillation index estimate, which would be applied to the frequency domain peak signal

$$\widehat{SI} = \sqrt{\left\langle\left|v\right|^4\right\rangle - \left\langle\left|v\right|^2\right\rangle^2} / \left\langle\left|v\right|^2\right\rangle,$$ (5.48)

where v is the signal-plus-noise model in $\sqrt{SNR}$ units. One can show by direct computation that

$$\widehat{SI} = \left(SI + 4/\left\langle SNR\right\rangle + 1/\left\langle SNR\right\rangle^2\right) / \left(1 + 6/\left\langle SNR\right\rangle + 1/\left\langle SNR\right\rangle^2\right),$$ (5.49)

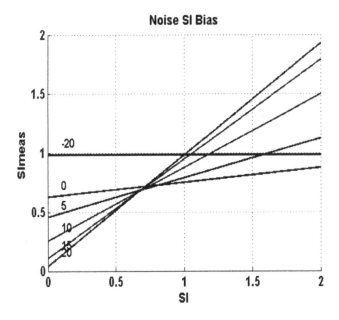

Figure 5.10 Measured scintillation index bias caused by additive white noise at the levels indicated by the SNR in dB.

where SI is the desired channel-induced signal scintillation index

$$SI = \sqrt{\left\langle |h|^4 \right\rangle - \left\langle |h|^2 \right\rangle^2} \Big/ \left\langle |h|^2 \right\rangle. \tag{5.50}$$

Figure 5.10 shows a plot of $\widehat{SI}$ versus SI as predicted by (5.49) for a range of $\langle SNR \rangle$ values. At zero SNR, $\widehat{SI} = 1$, the value of the SI index for the Gaussian noise background. To the extent that noise dominates the signal, the measured SNR value is pulled toward the unity noise value. This bias is negligible only for SNRs approaching 20 dB. If the SNR were known, a bias correction could be made. However, the uncertainty in estimating $\langle SNR \rangle$ at low SNR and the weak sensitivity of $\widehat{SI}$ to changes in SI make accurate SI extraction problematic. Even so, a well-designed system can readily achieve a processed SNR of 20 dB.

5.4.2 Digital Signal Processing

Modern digital signal processing operations generally start with the complex baseband receiver output. The means of acquiring this output is engineer's art perfected over many decades of technological progress. The heart of the operation is a phase-locked loop, which can take a variety of forms depending on the particular application. The analyst's luxury is to work with a digitally sampled signal from which the carrier has been removed. To illustrate the digital signal processing challenge, it is instructive to start with a simulated ideal signal. A symmetric, purely quadratic phase variation is not too far removed from the Doppler history of a high-elevation satellite track. With

$$s(n\Delta t) = A \exp\left\{i\frac{a}{2}\left(n\Delta t\right)^2\right\}, \qquad (5.51)$$

where A and a are parameters that set the signal strength and symmetric Doppler range, the instantaneous Doppler shift is obtained by converting the phase in radians to cycles and differentiating:

$$\begin{aligned} f_{Dop}\left(n\Delta t\right) &= 2\pi\frac{d\phi\left(t\right)}{dt}. \\ &= \pi a\left(n\Delta t\right). \end{aligned}$$

A realization of the complex signal is readily constructed as

$$v(n\Delta t) = \sqrt{SNR}\exp\left\{i\frac{a}{2}\left(n\Delta t\right)^2\right\} + (\zeta_n + i\zeta_n')/\sqrt{2}, \qquad (5.52)$$

where ζ_n and ζ_n' represent uncorrelated unit-variance zero-mean Gaussian variates. At this point there is no scintillation. Thus, the digital processor should recover an estimate of $s(t)$. A second luxury afforded the analyst is that the estimate usually does not have to be produced in real time.

For most applications the analyst can estimate the initial frequency. For example, orbit codes that tell the receiver when and where to look for a satellite pass can also predict the initial Doppler shift. However, with access to the recorded data, an initial frequency estimate can be generated with a Fourier transformation. To the extent that $\dot{\phi}(t)$ can be approximated as

$$\dot{\phi}(t) \simeq \dot{\phi}(t_n) + 2\pi f_{Dop}\left(n\Delta t\right)\Delta t, \qquad (5.53)$$

over the sample intervals $t_n < t < t_n + T$, a spectral density estimate will produce a narrow peak at $f_{Dop}\left(n\Delta t\right)$ Hz. The quadratic term introduces some smearing that depends on the frequency change during the coherent processing interval. A representative data segment was generated to present a -1000 to 1000 Hz sweep over $512 \times 10{,}000$ samples. At a 25 kHz sample rate, the coherent processing interval is $T = 512/25000 = 20.48$ ms.

Figure 5.11 shows the PSD derived from the first 512 samples of the realization. The noise level was adjusted to correspond to an input SNR of -7 dB. The measured frequency of the peak was 998 Hz, reflecting the small Doppler decrease from the initial 1 kHz frequency. The 20 dB processed SNR reflects ~ 27 dB processing gain achieved by the Fourier transformation. There is an additional important point to be made. The PSD estimate shown in Figure 5.11 used 32-fold oversampling. That is, the 512-sample complex signal

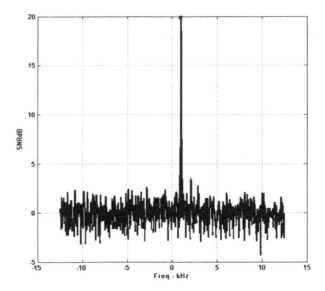

Figure 5.11   PSD of first 512-sample record from simulated data with 20 dB post-processed SNR. The signal peak is near 1 kHz.

vector was zero padded to $512 * 32 = 16384$ samples prior to computing the DFT. No windowing was used because sidelobe reduction, which is achieved at the expense of frequency resolution, is not needed for isolated peaks. The precision with which the frequency peak can be estimated is limited only by the SNR. High SNR is necessary not only for detection but also for accurate frequency estimation.

5.4.2.1   Doppler Tracking and Phase Extraction. The first objective of signal processing is to track the change in frequency of the signal peak. Simply repeating an unaided search for frequency peaks is inefficient and also risks confusion if there is more than one peak. For now, assume there is only one significant peak and it is close to $\overline{f}_{Dop}$. Let

$$\rho_N = \exp\left\{2\pi i\left[-N/2, -N/2 + 1, \cdots, N/2 - 1\right]/N\right\}, \tag{5.54}$$

where N is the number of complex data samples in the record. Forming the fractional power $\rho_N^{\overline{f}_{Dop}+\delta f}$ generates an ideal signal vector with a frequency near $\overline{f}_{Dop}$. Using the data vector $\mathbf{v}_N$ with conjugated phase, the magnitude of the dot product is a measure of the likelihood that the frequency hypothesis is correct

$$vH\left(\overline{f}_{Dop} + \delta f\right) = \left|\rho_N^{\overline{f}_{Dop}+\delta f} \cdot \mathbf{v}_N\right|^2. \tag{5.55}$$

This likelihood measure achieves a single maximum in a limited search range near the frequency hypothesis that matches the actual Doppler shift. The additive noise distorts the peak, which causes the estimate to fluctuate about the true value. An efficient algorithm that searches a narrow range of frequencies about $\overline{f}_{Dop}$ to find the best frequency match is readily constructed. Signal phase can then be constructed by integrating frequency. The trapezoidal rule

$$\phi_{n+1} = \phi_n + 2\pi \frac{\left(f_{n-1/2} + f_{n+1/2}\right)}{2} \Delta t_{\text{rec}} \tag{5.56}$$

accommodates the midpoint of the interval as the reference. The time interval between signal vector samples is Δt_{rec}.

5.4.2.2 Frequency and Phase Errors

To illustrate the performance of this recursive frequency tracking and phase estimation procedure, it was applied to a 10,000-record realization. The differences between the actual and measured frequency and derived phase estimates are shown, respectively, in the upper and lower frames of Figure 5.12. The Doppler errors are uncorrelated, whereas the phase errors show a low-frequency meander. This can be understood by the frequency-domain equivalence

$$
\begin{aligned}
I(t) &= \int_0^t f(t) \exp\left\{-i\omega t\right\} dt \\
&= \int \left[\frac{\widehat{f}(\omega)}{i\omega}\right]_\epsilon \exp\left\{i\omega t\right\} \frac{d\omega}{2\pi} + 2\pi \overline{f} t, \tag{5.57}
\end{aligned}
$$

where $[\cdots]_\epsilon$ indicates that the contribution about zero has been eliminated to avoid the singularity. The DC contribution is reintroduced explicitly with $\overline{f}$ representing the mean frequency obtained by integrating the function over the coherent processing interval. The low-frequency content is enhanced to the limit ϵ. To reduce the phase noise to less than 1 radian RMS requires an SNR greater than 30 dB, which further reinforces the high SNR requirement.

5.4.3 Multi-Frequency Data

Data from a Scion Associates multi-channel receiver developed for satellite scintillation measurements at 1067 (L-Band), 400 (UHF), and 150 (VHF) MHz were made available to illustrate digital signal processing and scintillation parameter estimation. The frequencies are transmitted by CERTO beacons carried by the C/NOFS satellite and the COSMIC constellation of 6 low Earth orbiting satellites. The Scion receivers coherently down-convert each signal by a factor of 10 where it is filtered, demodulated, and sampled at 25 kHz. The digital data are recorded as signed 16-bit integers interlaced in 3-channel by 512-sample records with UT tags. The data sample interval is $\Delta t = 40$ μs. The record repetition interval is $T = 20.48$ ms. A header record provides

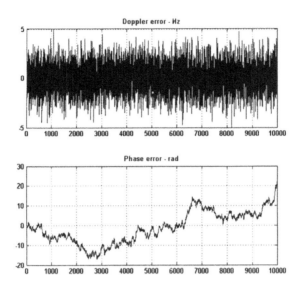

Figure 5.12   Doppler and phase errors with 20 dB post-processing SNR.

operational information including the receiver GPS coordinates and two-line orbital elements for each recorded pass.

Figure 5.13 shows a VHF power spectral density estimate obtained from a C/NOFS satellite pass recorded at an equatorial station.[19] The PSD estimate was obtained by averaging individual PSD estimates from the first 20 records. No windowing was used so that maximum frequency resolution is preserved. Non-coherent averaging is used to improve the noise passband estimate. Three narrowband signals are easily identified. An estimate of the receiver passband noise level has been overlaid on the PSD estimate. The noise estimate was obtained by recursively smoothing the PSD with outliers replaced by the local mean of the PSD. The real data present two variations from the ideal simulations. The noise is not white and there is more than one signal peak. The outer PSD peaks near ∓ 3 kHz can be identified as non-satellite signals because they do not change frequency with time.

The preprocessing operations are used only to estimate the noise background and to derive an initial estimate of the signal frequency. However, as seen in Figure 5.14, a zoom view of the 18 dB signal peak with 32 times oversampling, the peak can be localized to better than 5 Hz, which was also demonstrated by simulation.

---

[19]The Sao Luis station and the CNOS satellite were introduced in Chapter 4 to illustrate the prediction of scintillation structure.

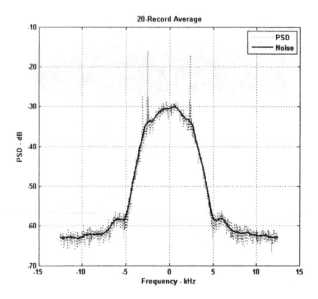

Figure 5.13   VHF PSD estimate obtained by averaging the raw PSDs from 20 consecutive records.

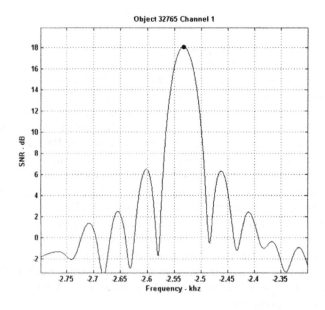

Figure 5.14   Zoomed-in view of 32× oversampled signal peak.

5.4.4 Frequency Tracking

The frequency tracking algorithm requires an initial frequency estimate for each channel. Although each channel could be processed independently, the directional gain of the VHF antenna is broader than the directional gains of the UHF and, particularly the L-band antennas. Consequently, the VHF signal has the highest SNR at the low elevations where the satellites are first acquired. Moreover, because the geometric Doppler dominates the signal Doppler shift, scaling the VHF Doppler estimate to UHF or L-band is sufficiently accurate to initiate Doppler peak searches in the auxiliary channels. For reference, Figure 5.15 shows the satellite range and range rate predicted by the SPG4 code described in Chapter 4 for the satellite pass being used for illustration here. The pass duration is ~ 7 minutes. Recalling that the geometric phase is $2\pi r/c$, and the corresponding Doppler shift is $-f_c\dot{r}/c$, it follows that the upper frame and the negative of the lower frame in Figure 5.15 are proportional to the geometric phase and Doppler shift, respectively. If the predictions were sufficiently accurate, these contributions could be removed analytically.

5.4.5 Signal Intensity

Because SNR drives the accuracy of all measurements it is prudent to start an analysis with signal intensity in SNR units. The upper and lower frames of Figure 5.16 show the VHF and UHF signal intensities, respectively. Each major UT time division in the display is 1.2 minutes. A 500-sample running mean estimate of SNR is overlaid on each plot. The VHF signal achieves a 20 dB SNR at the start of the pass. The UHF signal does not achieve a 20 dB SNR level for ~ 2.5 min. The variations of mean signal intensity are attributed mainly to the transmitter antenna patterns with fine structure caused by multipath from satellite body.

5.4.6 Signal Doppler

The corresponding Doppler estimates with UHF scaled by the VHF-to-UHF frequency ratio ($150/400 = 3/8$) are shown in Figure 5.17. On the kHz scale of the plot the differences are imperceptible, which confirms that the frequency tracking algorithm can capture a narrowband signal where the UHF SNR is very low. The VHF Doppler shift predicted by SPG4 is also shown in Figure 5.17. At zero Doppler, the predicted and measured values agree to the resolution limit, however, there are significant differences elsewhere. For comparison purposes, the predicted Doppler shift was offset to match the measured zero transitions. The offset curve has positive errors at both the high and low Doppler extrema, which implies significant quadratic error terms between the predicted and measured Doppler shifts. The SPG4 predictions

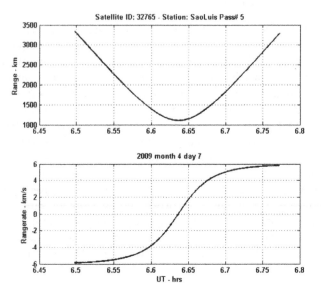

Figure 5.15   Satellite range and range rate predictions from SPG4 code using six-day-old orbital elements.

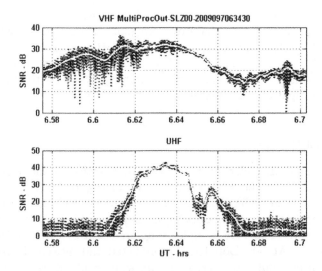

Figure 5.16   VHF and UHF intensity estimates from frequency tracking algorithm. A mean intensity estimate is superimposed for reference.

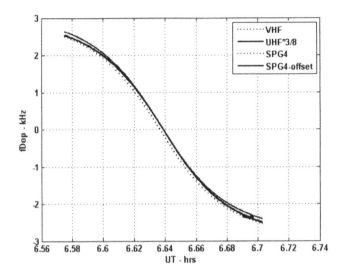

Figure 5.17   Measured and predicted Doppler shifts for VHF, UHF scaled, and SGP4 model.

are not designed to provide the wavelength level accuracy needed to remove the geometric Doppler contribution.

The VHF and UHF Doppler residuals derived by subtracting running mean estimates from the Doppler signals are shown in Figure 5.18.   From the beginning of the pass to 6.64 UT, both the VHF and UHF residuals follow the simulations with errors inversely proportional to SNR.   Toward the end of the pass, correlated Doppler noise bursts appear, which are unexplained. Since separate LNAs are used for each channel, the noise contributions should be uncorrelated.

To guide the interpretation of the phase data, scintillation theory predicts three distinct phase contributions,

$$\phi\left(t\right) = -2\pi r\left(t\right)/\lambda - 2r_e\lambda\overline{TEC} + \phi_{scint}\left(t; \lambda_1, r\right), \qquad (5.58)$$

where $\overline{TEC}$ is the path-integrated (total) electron density content (TEC), which is measured in electrons per unit area of integrated path distance.   Because the geometric contribution and the $\overline{TEC}$ occupy the same large-scale regime, it is appropriate to remove the small-scale structure, which is associated with noise and diffraction.   The time-honored method of separating the geometric phase contribution uses the scaled phase difference

$$
\begin{aligned}
\Delta\phi\left(t\right) &= \phi_{VHF}\left(t\right) - \left(\lambda_{UHF}/\lambda_{VHF}\right)\phi_{UHF}\left(t\right) \\
&= -2r_{e}\lambda_{VHF}\left(1 - \left(\lambda_{UHF}/\lambda_{VHF}\right)^{2}\right)\overline{TEC} - \Delta\phi_{scint}, \quad (5.59)
\end{aligned}
$$

which automatically removes the geometric Doppler but leaves a phase residual that potentially contains structure from each phase estimate. This introduces a scale separation between the $\overline{TEC}$ and the more rapidly fluctuating and smaller phase residuals. Figure 5.19 shows the phase difference defined by (5.59) in the upper frame with the mean overlaid. The lower frame shows the mean converted to TEC units.

5.5   SCINTILLATION DATA INTERPRETATION

In the development of the scintillation theory, scintillation was characterized as a modulation imparted to a signal that would otherwise propagate as if it were in a structure-free background environment. In the real world one must contend with intensity variations that for the most part cannot be removed analytically. Under the best conditions these variations come from smoothly varying antenna patterns. This variation can be minimized if the receiving antenna can track the source, but this type of operation is expensive to support. Similarly, if the satellite antenna is pointed optimally, for example, downward in a near circular orbit, a well-designed antenna can provide nearly uniform illumination over the central portion of the pass. Common phase centers and complementary patterns at all frequencies are also desired. One must also contend with limited frequency separation, which can compromise the measurement of phase scintillation. A large difference between the highest and lowest measurement frequencies is desired. Maintaining uniform illumination and a large frequency separation were guiding principles for the design of the Wideband Satellite beacon, which transmitted at S-band, L-band, UHF, and VHF [6]. Because satellite launches, even for modest payloads, are expensive and time-consuming to plan and execute, more recent and future beacon satellite experiments must share real estate and operating time with other experiments. As a consequence, there are some compromises that the analyst must contend with.

5.5.1   Scintillation Intensity Analysis

The results of design compromises are seen in the intensity estimates shown in Figure 5.16. The UHF system evidently has a very different geometric response than the VHF system. The L-band pattern is so narrow it is captured by the non-tracking receiving antenna only for short segments of nearly overhead passes. The intermediate scale structure is the most troublesome because the time scales involved are short enough that propagation sources

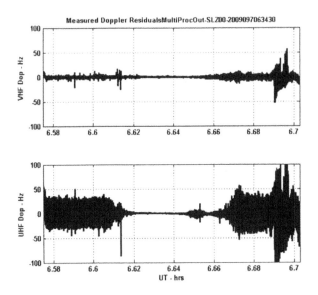

Figure 5.18   Doppler residuals obtaind by removing running mean estimate.

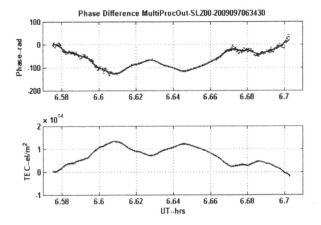

Figure 5.19   Smoothed phase difference from VHF and scaled UHF data (upper frame) with conversion to TEC units (lower frame).

might be of interest. The best one can do, however, is to chose a segmentation interval for mean signal estimation that follows the general intensity trend. The upper frames of Figures 5.20 and 5.21 show the detrended VHF and UHF intensity records, respectively. The detrended data are obtained simply by dividing the signal intensity by the running mean. In radar signal processing, the same operation is performed to maintain a constant false alarm ratio with a fixed threshold. To guide the interpretation of the data, however, the segments where the SNR exceeds 20 dB are marked on the -20 dB axis.

The lower frames of Figures 5.20 and 5.21 show the estimated SI index. In the initial portion of the pass where the scintillation activity reaches a moderate level, the VHF data in Figure 5.20 are not noise dominated and the measured index is reliable. Indeed, the VHF SI estimate near 0.8 can be reconciled with the SI model shown in Figure 5.10 for a nearly zero SNR. The result provides further evidence that the VHF queued tracking procedure successfully finds a signal in the noise-limited UHF data. Unfortunately, the region where the SI estimate reflects the signal is almost disjoint with the moderate scintillation at VHF. This occurs only because the UHF antenna response is much narrower than the VHF antenna response.

5.5.2 Spectral Analysis

Spectral representations were used extensively in Chapters 2, 3, and 4 to relate scintillation structure to the media structure that initiates the process. In Chapter 4 it was shown that spectral analysis of simulated data can provide robust measures of the in situ structure, even when the evolving scintillation structure is not homogeneous within the measurement frame. However, this analysis addresses small-scale structure that can be regarded as statistically homogeneous. The single data set introduced here to demonstrate scintillation preprocessing is too small and limited to fully demonstrate the spectral analysis procedures in detail, but some conclusions can be drawn.

By way of review, TEC measurement as represented in the lower frame of Figure 5.19 is the most robust data product. Using the phase residual that was eliminated prior to forming the TEC estimate as a measure of phase scintillation is problematic for two reasons. The small-scale structure is compromised by low SNR except in the central portion of the pass. Also, to the extent that geometric phase change can be removed only by scaled phase differences, the UHF frequency is too close to the VHF frequency to separate the diffraction effects. As a result, there is no viable way to extend the phase structure to smaller scales than those that can be interpreted as TEC variations. An investigation by Nickisch [40] based on spectral analysis of TEC data has already been noted in Chapter 3.

Turning to intensity scintillation, detrending was introduced as a viable means of removing geometric variations introduced mainly by antenna patterns. Unfortunately, scintillation data at moderate SI levels were only ob-

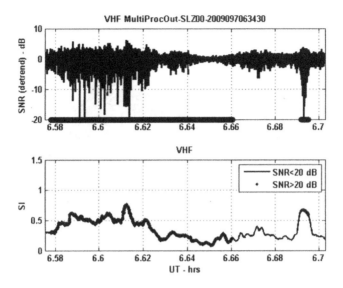

Figure 5.20   Detrended VHF intensity (upper frame) and SI index (lower frame) from processed Scion data. Segments with SNR>20 dB are marked.

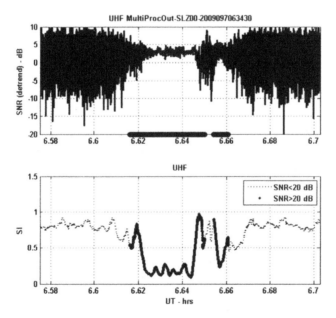

Figure 5.21   Detrended UHF intensity (upper frame) and SI index (lower frame) from processed Scion data. Segments with SNR>20 dB are marked.

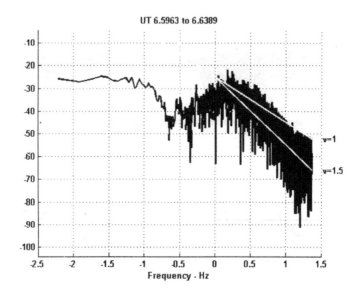

Figure 5.22   Spectral analysis of detrended intensity data from a single data segment capturing the initial large SI region.

tained at VHF during the initial portion of the pass. In Figure 5.20, data with SNR>20 dB were processed to extract SI estimates over fairly short segments. Following the analysis presented in Section 4.6, a much larger processing interval is desired. Figure 5.22 shows the computed intensity PSD and power-law spectra with $p = 2\nu$ overlaid with slopes for $\nu = 1$ and $\nu = 1.5$. The initial variation of the intensity spectra is a residual of the detrending operation. The spectral peak near 0.25 Hz is attributed to Fresnel filtering. The f^{-3} power-law behavior is typical.

Additional discussion on this topic can be found in the review paper by Bhattacharyya, Yeh, and Franke [107], and the second volume of *Electromagnetic Scintillation* by Wheelon [108]. The subject is an active area of research. It has already been noted in Section 4.5.3 that the estimation of power-law parameters is itself an area of study. Because of the complexity introduced by the varied ionospheric structure sources and the critical dependencies on observational geometry, organizing data is particularly challenging.

5.6 BEACON SATELLITE RESEARCH

Interest in the theory of scintillation and its applications to remote sensing is often motivated by what can be learned from scintillation observations. The development in this book concentrated on the relationship between observations and structure that facilitates the process. However, it is appropriate to conclude with some background, an overview of current research, and some future opportunities to which the theory can be applied.

Beacon satellite observations started impulsively with the launch of Sputnik in 1957. The early development is described in review papers by Aarons [2], Crane [109], Yeh and Liu [1], and more recently by Aarons [110]. This recent review covers the early observations chronologically. The first satellite beacon designed for ionospheric measurements was part of the Applied Technology Satellite (ATS) series of synchronous equatorial satellites [39]. Satellite navigation also gave ionospheric scintillation studies a boost. Tracking space objects to determine their orbits has a long history in observational astronomy; however, with an active beacon from which the Doppler shift induced by the source motion can be measured, it was recognized that the problem could be reversed. If the satellite broadcast its position, Doppler tracking then could be used to determine the observer's position very accurately. This principle was exploited by the U. S. Navy to solve a critical covert submarine navigation problem. The Navy Navigation Satellites (NNS) provided extremely reliable coherent 150 and 400 MHz beacon transmissions that were visible over most of the Earth.

The high reliability of the NNS TRIAD satellite payloads and their Scout D launch vehicles led to an excess inventory that inadvertently gave an additional boost to radio beacon research. Three of the NNS payloads were reconfigured and launched as dedicated radio beacon resources. The first was the Wideband satellite, launched from Vandenburg Air Force Base on 22 May 1976. The second was the Hilat satellite launched 27 June 1983. The third was the Polar Bear satellite launched in November 1986. As the names imply, Hilat and Polar Bear were designed for Auroral Zone observations under magnetically disturbed conditions. Instrumentation to measure optical emissions from auroral activity were included in the payloads.

As technology advanced further, satellite systems moved to higher frequencies. The GPS satellites transmitted dual frequency L-band signals that could be processed to extract intensity scintillation and ionospheric TEC. These satellites were also used for atmospheric scintillation measurement and, more important for atmospheric occultation experiments that could be inverted to estimate refractivity profiles. The utility of multiple widely spaced receivers for tomographic inversion of large-scale ionospheric structure was developed and refined as part of a broad range of new GPS applications.

Radio astronomers, in the meantime, were pioneering novel applications of propagation phenomenology that include large baseline interferometry, planetary radio occultation measurements, and solar wind monitoring. Novel new applications of GPS led to the COSMIC science mission, which exploits the radio occultation technique using the constellation of GPS satellites as sources and low Earth orbiting (LEO) satellite receivers to monitor the Earth's atmosphere [111].   The potential for improved ionospheric scintillation measurement was recognized and exploited.   The Coherent Electromagnetic Radio Tomography (CERTO) transmitter was designed for extensive ionospheric measurement [112], particularly computerized ionospheric tomography (CIT). The CERTO beacon was added to the U.S. Air Force Communications Outage Forecast System C/NOFS, which was the data source for the example presented in this chapter.   The Air Force system works in conjunction with C/NOFS, which is intended to forecast propagation conditions based on all available information.   The Scintillation and Tomography Receiver in Space (CITRIS) was designed for both ground- and space-based operation.   The results that are beginning to emerge ([113], [114]) show the potential of multisatellite observations that exploit geometry and complementary observations.

Global atmospheric monitoring has taken on a new importance with concerns about how global warming is changing our environment.   The main concern is the Earth's atmosphere, which is measured directly through refraction with the COSMIC LEO satellites.   Nonetheless, the ionosphere does have the potential to degrade and even disrupt GPS under disturbed conditions.   Additionally, the structure that develops in the Earth's ionosphere is an accessible plasma laboratory that should not be overlooked.   *The Earth's Ionosphere* [25], now in its second edition, provides an up-to-date compact treatment of the physics of global ionospheric dynamics and structure development.   Radar measurements and in situ probes provided most of the data, but scintillation remains as the primary diagnostic for quantitative measurement of intermediate-to-large-scale structure.

CHAPTER 6

SCATTERING AND BOUNDARIES

Everything should be made as simple as possible, but not simpler.

—Albert Einstein

The scintillation theory developed and explored in Chapters 2 through 5 was restricted to unbounded weakly inhomogeneous media. In real-world applications, however, scatter from compact objects and surfaces is difficult to avoid. More importantly, most remote sensing applications rely on boundary scattering. Boundary scattering theory is used most often to model scattering from compact objects and surfaces, but it is complicated and computationally demanding in its most general form. The source of this complexity is readily demonstrated. Consider a collection of N scattering centers. When a wave field interacts with one of the scattering centers, the scattered wavefield potentially illuminates all N scattering centers. Thus, the incident field at each scattering center, which is called the excitation field, is the sum of the direct wave plus the fields initiated at every other scattering center. A self-consistent calculation of the N unknown excitation fields requires the solution of N linear integral equations [115]. In boundary scattering theory, a continuum of discrete scattering centers supports surface currents that sat-

The Theory of Scintillation with Applications in Remote Sensing, by Charles L. Rino **177**
Copyright © 2011 Institute of Electrical and Electronics Engineers

isfy surface boundary conditions. The resulting boundary integral equations (BIEs) are solved by replacing the boundary surface with a discrete set of scattering centers, which is structurally identical to the discrete scatter problem. It is possible to reduce the N^2 growth in complexity to $N \log N$, but direct numerical solutions to large-scale scattering problems remain formidable. This chapter will show that the forward approximation can be exploited to obtain more tractable solutions to problems that include compact scattering objects embedded in weakly inhomogeneous media bounded by irregular surfaces.

Embedded compact scattering objects are treated first. To the extent that structured layers and known scattering objects can be separated, the mutual interaction problem admits an exact solution. The double-passage radar problem in which a compact target is viewed through an intervening structured layer is a special case. Because of its practical utility, the double-passage problem is treated in detail. The development shows how a well-known backscatter radar cross section (RCS) enhancement develops. It also establishes a reciprocity relation that can be used to avoid explicit calculation of the backward propagation step.

An irregular boundary presents a much more challenging problem. If the scattering characteristics of a boundary layer can be established independently, the same technique that was used for the double-passage problem can be applied to calculate the backscatter from the layer through an intervening disturbed layer or from a compact object near the surface [16]. It is important to note as well that the coordinate system for this type of surface scattering problem uses the axis normal to the surface reference plane as the propagation reference axis. Indeed, the Bragg wavenumber that supports Bragg surface scatter applies to a two-dimensional spectral decomposition of the structure in a horizontal plane. The theory breaks down when the incident wave direction approaches grazing incidence.[20] In the scintillation framework this would represent propagation along the layer rather than through the layer. The real-world limit imposed on grazing incidence is readily demonstrated. To resolve a very small scattering angle requires a very large illuminated active area. Thus, as grazing incidence is approached, the active support that influences the propagation increases without bound. Under these conditions, the problem transitions from a scattering problem to a guiding surface problem. The formulation is greatly simplified by changing the propagation reference from normal to the surface to parallel to the surface.

For propagation in a bounded weakly inhomogeneous medium, the problem can be approached by using conventional differential equation methods. Canonical boundary conditions require specification of the field and its normal derivative at the boundary. At this point a reader familiar with PWE methods might claim the problem has been solved. Indeed, for cylindrical wave (two-dimensional) propagation over a plane surface, the sine and cosine

---

[20]Grazing angle is the complement of the incidence angle, which is measured from the principal propagation axis.

transform components of the propagator intrinsically satisfy one-dimensional conducting boundary conditions. The electric field is either parallel to the surface (horizontal polarization) or in the propagation plane (vertical polarization). There is no polarization coupling. The fact that the Fourier-domain propagator can be decomposed into sine and cosine components leads directly to a forward-marching solutions that readily accommodates weakly inhomogeneous structure. One can also construct a combined boundary condition that accommodates impedance boundary surfaces[116]. Although these results are strictly applicable only for plane surfaces, conformal calculations in which the irregular surface defines the vertical origin have been employed effectively. The applications of parabolic-wave-equation methods to this type of problem are reviewed in detail in the book *Parabolic Equation Methods for Electromagnetic Wave Propagation* [30].

The approach pursued here, by comparison, is a direct application of the FPE formalism to the boundary-integral theory of the surface scattering problem. The approach is attractive as a more rigorous alternative to the PWE methods that can be extended to fully three-dimensional problems. The computational requirements are severe and the implementation is demanding. As a consequence, the unrestricted theory is developed only in algorithmic form. The examples that illustrate its implementation are restricted to two-dimensional propagation. The complete two-dimensional BIE theory is summarized in Appendix A.5. An impedance boundary condition with a surface curvature correction is included. (See Tables 6.1 and 6.2 for symbols and abbreviations.)

Table 6.1   Chapter 6 Symbols

Symbol	Definition
$\widehat{\psi}^{\pm}(x_n, \boldsymbol{\kappa})$	Wave spectrum at forward (+) or trailing boundary plane (-)
$S_n^{\pm\mp}(\boldsymbol{\kappa}, \boldsymbol{\kappa}')$	Scattering function for nth layer
$P_n^u(\boldsymbol{\kappa})$	Propagation operator (6.2)
$S_n^{u\pm}@\widehat{\psi}_{n-1}^{\pm} \Big|_n$	Scattering interactions (6.3)
$\Phi(\boldsymbol{\kappa}, \boldsymbol{\kappa}_i)$	Scattered intensity spectrum (6.20)
$\delta s_m(f)$	Point frequency-dependent radar scattering model
s_m^n	Preprocessed range (n)/Doppler (m) response (6.26)
$\overline{\Gamma}(\mathbf{r}, \mathbf{r}')$	Dyadic Green function (6.30)

Table 6.2   Chapter 6 Abbreviations

Abbreviation	Definition
BIE	Boundary integral equation
MFIE	Magnetic field integral equation
EFIE	Electric field integral equation
MOMI	Method of ordered multiple interactions

6.1 EMBEDDED COMPACT SCATTERING OBJECTS

Solutions to the FPE characterize fields that propagate only in the forward (positive x) direction. Following the development in Rino [117], the scattering system to be solved is comprised of discrete scatters and structured layers that can be isolated by planes located at $x = x_n$. At each plane the total field is decomposed into two-dimensional plane-wave spectra denoted $\widehat{\psi}^{\pm}(x_n, \boldsymbol{\kappa})$. The superscript indicates the propagation direction. Although the weakly homogeneous layers do not directly support backscatter, backscattered energy from the scattering objects provides a source of backward propagating but weakly interacting waves. The more compact notation $\widehat{\psi}_n^{\pm}(\boldsymbol{\kappa}) = \widehat{\psi}^{\pm}(x_n, \boldsymbol{\kappa})$ will be used in the development. Both scattering layers and scattering objects can be characterized by a plane wave-to-plane wave bistatic scattering function $S^{\pm\mp}(\kappa, \kappa')$. For a scattering layer, the scattering function is the spatial Fourier domain solution to the FPE, which has no backscatter contribution. The signs refer respectively to the incident and scattered reference propagation directions relative to a fictitious plane through the phase center of the scattering object or layer. Only co-polarized (scalar) interactions will be treated explicitly, although the methodology can be extended to a full vector treatment.

6.1.1 Mutual Interaction Formulation

To set up the mutual interaction problem, the scattering entity characterized by $S_n^{\pm\mp}(\kappa, \kappa')$ is isolated by planes at x_{n-1} and x_n, with l_n^- and l_n^+ representing the propagation distances from x_{n-1} to the phase center and from the phase center to x_n, respectively. The scattered plane wave components observed at the boundary planes are computed by appropriate integrations over the incident plane wave fields. The following compact notation is used:

$$\widehat{\varphi}_{n-1 \atop n}^u(\kappa) = P_n^u(\kappa) \exp\left\{i(\kappa - \kappa') \cdot \varsigma_n\right\} S_n^{u\pm}(\kappa, \kappa') P_n^\mp(\kappa'), \tag{6.1}$$

where u identifies the direction of the incident wave. The upper and lower indices are used with the corresponding upper and lower signs. The spectral-domain propagator is defined as

$$P_n^u(\kappa) = \exp\left\{ik_x(\kappa) l_n^u\right\}, \tag{6.2}$$

where ς_n defines the location of the scattering center within the phase center plane. The following compact notation is used in the formulation of the scattering interaction when the excitation field is a continuous spectrum of plane waves rather than a single plane wave:

$$
\begin{aligned}
S_n^{u\pm} @ \widehat{\psi}_{n-1 \atop n}^\pm &= P_n^u(\kappa) \iint \exp\left\{i(\kappa - \kappa') \cdot \varsigma_n\right\} S_n^{u\pm}(\kappa, \kappa') \\
&\quad \times \widehat{\psi}_{n-1 \atop n}^\pm(\kappa') P_n^\mp(\kappa') \frac{d\kappa'}{(2\pi)^2},
\end{aligned} \tag{6.3}
$$

The energy conservation and reciprocity properties of wave wave scattering functions were discussed in Section 1.1.2 and the references cited. Reciprocity is an explicit property of Green functions, which characterize propagation from point sources. The corresponding reciprocity relations for spectral-domain scattering functions $S_n^{u\pm}(\kappa, \kappa')$ are

$$S_n^{\pm\pm}(\kappa, \kappa')/g(\kappa') = S_n^{\mp\mp}(-\kappa', -\kappa)/g(\kappa) \tag{6.4}$$

$$S_n^{\pm\mp}(\kappa, \kappa')/g(\kappa') = S_n^{\mp\pm}(-\kappa', -\kappa)/g(\kappa) \tag{6.5}$$

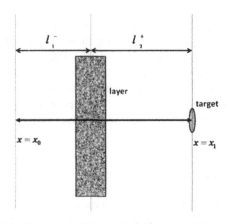

Figure 6.1   Schematic representation of the double-passage geometry.

For an isotropic point scatterer with cross section σ,

$$S_n^{\pm\mp}\left(\kappa, \kappa'\right) = \frac{(\pi\sigma)^{1/2}}{kg(\kappa)}. \tag{6.6}$$

Note that $S_n^{\pm\mp}\left(\kappa, \kappa'\right) / g\left(\kappa'\right)$ satisfies the reciprocity relations.

6.1.2 Double-Passage Propagation

To set up a system of equations that can be solved for the unknown fields at each boundary plane, it is necessary to use only (6.3) to characterize the fields crossing the boundary planes. The interaction relations developed in the previous section are used. The incident field at the first plane initiates the problem. If multiple interacting scatterers lie within a layer, a combined scattering function can be evaluated to construct a composite scattering function. Detailed discussions of the method and its applications can be found in the references [13], [14], and [15].

The double-passage problem is formally represented as a two-component scattering system consisting of a turbulent layer between planes at $x = x_0$ and $x = x_1$ and an isotropic scatterer located at the exit plane $x = x_1$. The configuration is illustrated schematically in Figure 6.1. The total wave spectrum generated by an arbitrary incident wavefield, $\widehat{\psi}_0^+\left(\kappa\right)$, can be written formally as

$$\widehat{\psi}_1^+\left(\kappa\right) = \widehat{\psi}_0^+\left(\kappa\right)\exp\left\{ik_x\left(\kappa\right)l_1\right\} + S_1^{++}@\widehat{\psi}_0^+, \tag{6.7}$$

where $l_1 = l_1^- + l_1^+$ and $S_1^{++}(\kappa, \kappa')$ is the forward scattering function for the layer. Proceeding with the formal development, the scattered field propagating back into the scattering layer can then be written as

$$\widehat{\psi}_1^-(\kappa) = S_2^{-+}@\widehat{\psi}_1^+, \qquad (6.8)$$

and the desired backscattered wave at $x = x_0$ is

$$\widehat{\psi}_0^-(\kappa) = \widehat{\psi}_1^-(\kappa) \exp\{ik_x(\kappa)\, l_1\} + S_1^{--}@\widehat{\psi}_1^-. \qquad (6.9)$$

By using reciprocity S_1^{--} can be derived from S_1^{++}. Thus, the forward scattering function, the scattering function for the object, and the source field provide all the information needed to calculate the backscatter at the source plane.

The symmetry between (6.7) and (6.9) suggests definition of a new scattering function, identified by an overbar, that characterizes the total field:

$$\begin{aligned}
P_n^u(\kappa)\, \overline{S}_n^{u\pm}(\kappa, \kappa')\, P_n^{\mp}(\kappa') &= P_n^u(\kappa)\, \delta(\kappa - \kappa')\, (2\pi)^2\, P_n^{\mp}(\kappa') \\
&\quad + P_n^u(\kappa)\, S_n^{u\pm}(\kappa, \kappa')\, P_n^{\mp}(\kappa'). \qquad (6.10)
\end{aligned}$$

The forward and backward equations then take the simpler forms

$$\widehat{\psi}_1^+(\kappa) = \overline{S}_1^{++}@\widehat{\psi}_0^+, \qquad (6.11)$$

and

$$\widehat{\psi}_0^-(\kappa) = \overline{S}_1^{--}@\widehat{\psi}_1^-. \qquad (6.12)$$

By substituting (6.11) into (6.8) and then substituting that result into (6.12), the following general solution to the double-passage problem is obtained for any excitation field and any compact scattering object:

$$\widehat{\psi}_0^-(\kappa) = \overline{S}_1^{--}@\left[S_2^{-+}@\left[\overline{S}_1^{++}@\widehat{\psi}_0^+\right]\right]. \qquad (6.13)$$

As already noted, the scattering functions for the total field, $\overline{S}_1^{\pm\pm}$, are Fourier-domain solutions to the FPE.

For an incident plane wave at $x = x_0$, (6.8) simplifies to

$$\widehat{\psi}_1^+(\kappa) = \overline{S}_1^{++}(\kappa, \kappa_i). \qquad (6.14)$$

If the scattering object is isotropic, then

$$\begin{aligned}
\widehat{\psi}_1^-(\kappa) &= S_2^{-+}@\overline{S}_1^{++}(\kappa, \kappa_i) \\
&= \frac{(\pi\sigma)^{1/2}}{kg(\kappa)} \iint \left[P_1^-(\kappa')\, \overline{S}_1^{++}(\kappa', \kappa_i)\, P_1^+(\kappa_i)\right] \frac{d\kappa'}{(2\pi)^2}. \qquad (6.15)
\end{aligned}$$

Substitution of (6.14) and (6.15) into (6.13) leads to the integral product form:

$$\widehat{\psi}_0^- (\kappa) = S_1^{--} @ \widehat{\psi}_1^-$$

$$= \frac{(\pi\sigma)^{1/2}}{kg(\kappa)} \iint P_1^- (\kappa) \frac{S_1^{--} (\kappa, \kappa')}{g(\kappa')} P_1^+ (\kappa') \frac{d\kappa'}{(2\pi)^2}$$

$$\times \iint \left[P_1^- (\kappa'') \overline{S}_1^{++} (\kappa'', \kappa_i) P_1^+ (\kappa_i) \right] \frac{d\kappa''}{(2\pi)^2}. \qquad (6.16)$$

To interpret the integral terms in (6.16), note that

$$\overline{S}_n^{\pm\pm} (x, \rho; \kappa) = \iint \overline{S}_{in}^{\pm\pm} (\kappa'; \kappa) \exp \left\{ \pm ik_x (\kappa') \left(x - x_{n-1} \pm l_{n-1}^\pm \right) \right\}$$

$$\times \exp \{ i\kappa' \cdot \rho \} \frac{d\kappa'}{(2\pi)^2} \qquad (6.17)$$

represents the field at (x, ρ) due to an incident plane wave from the direction κ consistent with the signs of the scattering function. It follows with an application of reciprocity that

$$\widehat{\psi}_0^- (\kappa) = \frac{(\pi\sigma)^{1/2}}{kg(\kappa)} P_1^- (\kappa) \overline{S}_1^{++} (x_1, 0; -\kappa) \overline{S}_1^{++} (x_1, 0; \kappa_i) P_1^+ (\kappa_i), \qquad (6.18)$$

and moreover that

$$\overline{S}_1^{-+} (\kappa, \kappa_i) = \frac{(\pi\sigma)^{1/2}}{kg(\kappa)} \overline{S}_1^{++} (x_1, 0; -\kappa) \overline{S}_1^{++} (x_1, 0; \kappa_i) \qquad (6.19)$$

is the composite scattering function that characterizes the backscattered plane wave spectrum for an incident plane wave with transverse wavenumber κ_i.

Now consider the field that would be measured at a long distance from the turbulent layer. Using the far field scaling described in Section 1.1, the result is

$$\Phi (\kappa, \kappa_i) = \frac{k^2 g(\kappa)}{\pi} \left\langle \left| \widehat{\psi}_0^- (\kappa) \right|^2 \right\rangle$$

$$= \sigma \left\langle \left| \overline{S}_1^{++} (x_1, 0; -\kappa) \right|^2 \left| \overline{S}_1^{++} (x_1, 0; \kappa_i) \right|^2 \right\rangle. \qquad (6.20)$$

It follows from (6.20) that when $\kappa = -\kappa_i$ the term in angle brackets is the second moment of the field intensity. The backscatter RCS enhancement result,

$$\Phi (-\kappa_i, \kappa_i) = \sigma \left(SI^2 + 1 \right), \qquad (6.21)$$

follows from the defining relation of the scintillation index. The reciprocal form of (6.21) also establishes the fact that the square of the point scatter from

from a one-way realization properly accounts for the two-way propagation. Calculating the fields with the spherical wave correction factor discussed in Chapter 4 shows that an identical result applies to two-way paths corrected for spherical-wave propagation. Formally, (6.20) is a measure of the spatial wavenumber decorrelation of intensity. However, other than the backscatter result (6.21), there is no analytic result akin to the intensity SDF derived in Chapter 3.

6.1.3 Radar Imaging Through Disturbed Media

The double-passage problem for a point scatterer sets the stage for the more general radar imaging problem. The first step is to apply the channel modeling framework developed in Chapter 5. Reciprocal two-way propagation to and from a source can be constructed from an FPE solution with plane-wave excitation. Introducing the CDCS geometry (from Chapter 4) and the waveform modulation $\widehat{v}_m(f, d)$ (from Chapter 5) leads to the signal model for a scattering center at location ρ:

$$\delta \widehat{s}_m^l(f) = A A_l(\rho)\widehat{v}_m(f, d)\psi_{\mathbf{k}}^2(\rho + \mathbf{v_k}t; \mathbf{R}_0, f_c + f)\sigma(\rho) + \widehat{\varsigma}(f). \qquad (6.22)$$

Strictly speaking, the scattering strength $\sigma(\rho)$ depends on both the polarization and direction of the incident wave. For now, assume that these factors have been absorbed in the radar reflectivity $\sigma(\rho)$. The noise contribution $\widehat{\varsigma}(f)$, the FPE solution $\psi_{\mathbf{k}}$, and the modulation $\widehat{v}_m(f, d)$ have already been defined. The two-way antenna amplitude gain $A A_l(\rho)$ can take a variety of forms depending on the configuration. With a single antenna used for both transmission and reception, $A A_l(\rho)$ can be replaced by $A^2(\rho)$, where $A(\rho)$ is the normalized antenna amplitude gain function. A single element carries a phase factor that accommodates its displacement from the phase center. Antenna arrays may be fully populated with active elements or, more often, with a small number of illuminators and a single fully populated larger receiving array. Practical constraints dictate the size, the configuration, and the number of elements for transmission and reception. In all cases, however, one can construct a per-element two-way amplitude response function $A A_l(\rho)$. Processing for the elements represented by $A A_l(\rho)$ will be discussed below. For now, it is sufficient to note that the transmit antenna must illuminate the measurement plane area of interest. The wave vector $\mathbf{k}$ indicates the direction of the illuminating wave field.

For a collection of non-interacting scattering centers, the contributions from each scattering element are summed to form the frequency-dependent waveform element response,

$$\widehat{s}_m^l(f) = \iint A A_l(\rho)\widehat{v}_m(f, d)\psi_{\mathbf{k}}^2(\rho + \mathbf{v_k}mT; \mathbf{R}_0, f_c + f)\sigma(\rho)\, d\rho + \widehat{\varsigma}_m^l(f). \qquad (6.23)$$

Each point in the measurement plane has an associated range $r(\rho)$ and range rate $\dot{r}(\rho)$ dictated by the source and platform dynamics. The Doppler shift

d is a function of the measurement plane position determined by $\dot{r}(\rho)$. Now consider the signal preprocessing operations described in Chapter 5. The receiver captures the delay response from each waveform, which is formally the inverse Fourier transform of (6.22),

$$s_{nm}^l = \int \tilde{s}_m^l(f) \exp\{2\pi i f n \delta t\} df.$$

To the extent that the channel supports the full bandwidth of the radar waveform, the effect of a matched filter with compensation as needed for Doppler offsets and range walk can be represented as follows

$$\hat{v}_m(f, d) \leftarrow P_s(\tau_n - 2r(\rho)/c) \exp\{2\pi i m (f_c T/c) \dot{r}\}. \tag{6.24}$$

The function $P_s(\tau_n - 2r(\rho)/c)$ is the intensity response of the matched filter, which is peaked about the nominal delay $\tau_n = 2r(\rho)/c$. The result is a processed frame with range by pulse number coordinates

$$\begin{aligned} s_{nm}^l &= \iint AA_l(\rho)P_s(\tau_n - 2r(\rho)/c) \exp\{2\pi i m (f_c T/c) \dot{r}\} \\ &\times \psi_{\mathbf{k}}^2(\rho + \mathbf{v_k} mT; \mathbf{R}_0, f_c)\sigma(\rho) d\rho + \varsigma_m^l. \end{aligned} \tag{6.25}$$

Similarly, the effect of applying Doppler processing to a sequence of range-resolved returns from M waveforms introduces $D_s(d_l - 2f_c \dot{r}(\rho)/c)$, a nonnegative function concentrated about $d_l - 2f_c \dot{r}(\rho)/c$. The result is processed frame with range by range-rate coordinates

$$\begin{aligned} s_{nk}^l &= \iint A_l^2(\rho)P_s(\tau_n - 2r(\rho)/c)D_s(d_k - 2f_c \dot{r}(\rho)/c) \\ &\times \psi_{\mathbf{k}}^2(\rho + \mathbf{v_k} n_{CPI}CPI; \mathbf{R}_0, f_c)\sigma(\rho) d\rho + \varsigma_{mk}^l. \end{aligned} \tag{6.26}$$

To summarize, each radar antenna element presents the signal processor with a data frame indexed by fast time and waveform number (slow time). The baseband frequency spectrum is represented by (6.23). After a matched filter operation has been applied to each waveform, the fast time-by-slow time data frame is converted to a data frame indexed by delay and pulse number. Doppler processing over the slow-time index converts the pulse-compressed data frame to range by range rate. To the extent that propagation induces no change over the waveforms that comprise the coherent processing interval, the result can be written as

$$s_{nk}^l \simeq AA^l(\rho^{nk})P_s(0)D_s(0)\psi_{\mathbf{k}}^2(\rho^{nk} + \mathbf{v_k} n_{CPI}CPI; \mathbf{R}_0, f_c)\tilde{\sigma}(\rho^{nk}) \Delta\rho + \eta_{nk}^l. \tag{6.27}$$

The index n is the range sample index, k is the range-rate index, and l is the antenna element or pointing direction index. The notation ρ^{nk} means the measurement-space location that corresponds to the nominal range and range rate associated with the corresponding index values. The channel-induced modulation varies with time only from CPI to CPI, which is indexed by n_{CPI}. Because the delay and Doppler response functions are highly concentrated, the usual interpretation of (6.26) is the radar equivalent of the slow-fading channel. This is one of the fading cases treated by Swerling as discussed in Chapter 4, but the source of the fading is usually the changing aspect of the target itself.

How and why cross-section measurements vary is an important topic. For an isolated compact target, (6.27) is essentially exact. Indeed, it provides the basis for verifying the cross-section enhancement predicted by (6.21) [118]. A calibration sphere was used to eliminate target-induced fluctuations. The purpose of a calibration sphere is to convert $\left\langle \left| s^l_{nk} \right|^2 \right\rangle$ to RCS units. The interpretation of the result for extended scattering structures and propagation environments remains challenging. Even in the absence of propagation disturbances, the localized radar returns as implied by (6.27) cannot be interpreted as the surface RCS. For example, one does not know how the resolved cell was illuminated. Through shadowing, a resolved cell may not be illuminated at all. Similarly, the cell may be illuminated by multiple reflections from different parts of the scattering structure. Finally, to the extent that the scattering object is an extended surface, the same data-space location can produce very different cross-section values when viewed from different aspects. All of these problems are well know in radar processing. The book *Fundamentals of Radar Signal Processing* [103], already cited in Chapter 5, presents a thorough topic-by-topic treatment.

For completeness, the possibility of adaptive processing should be mentioned. The standard use of element-by-element processing is beam forming. Where jamming and clutter must be dealt with, one can adaptively measure the undesired inputs. To the extent that these inputs reside in parts of the data space that can be isolated, adaptive processing can eliminate the unwanted signal inputs. Since jammers and clutter are typically much stronger that the desired signal, a high degree of cancellation is required. The degree to which this cancellation can be achieved is acutely sensitive to deviations from the ideal signal vectors [119]. Adaptive processing can be applied earlier in the processing sequence as well. Simulations based on (6.22) are ideally suited for establishing performance bounds on what can be achieved.

6.1.4 SAR Example

Synthetic aperture radar (SAR) and its counterpart, inverse synthetic aperture radar (ISAR), formally exploit a relation between radar coordinates and data space coordinates. The functional relation between the data space position vector ρ and the radar coordinates $r(\rho)$ and $\dot{r}(\rho)$ was established in the development of (6.23) and (6.26). Indeed, to the extent that ρ is confined to the boundary surface of a scattering object, the surface can be referenced by radar coordinates. The mapping from surface position to radar coordinates usually is not one-to-one, but antenna directivity and the choice of platform motion relative to the surface can be used to eliminate the most troublesome ambiguities. To the extent that range and range rate are being measured on a *known* surface, the SAR and ISAR imaging problems reduce to inverse mapping.

The degree to which the surface is known a priori separates SAR from ISAR processing. SAR processing identifies structure on a surface whose position and orientation are known reasonably well. ISAR processing first locates the centroid of the scattering object. Rotation hypotheses are then used to find an inverse mapping that focuses the image. This conceptual view based on coordinate mapping is more general than the more familiar description of SAR imaging as forming beams over an aperture constructed synthetically as the radar data are collected. Where the mapping is one-to-one, the two interpretations are fully consistent.

To illustrate the inverse-mapping interpretation of SAR processing, consider an aircraft flying at 400 kts while maintaining a constant altitude of 20,000 ft. An antenna on the side of the aircraft has its beam pointed downward at 60° from nadir. Figure 6.2 shows a random image plane intensity map that represents $\sigma(\rho)$. In the absence of propagation disturbances, a processed radar frame would capture an approximation to $\sigma(\rho)$ on the range-azimuth grid overlaid on the image in Figure 6.2. The horizontal arcs are contours of constant range. The nearly vertical lines are contours of constant range rate. To generate a representative radar image after preprocessing to isolate the radar returns, the realspace coordinates are mapped to range and range-rate cells and integrated over the nominal resolution cell. A more accurate reconstruction of an ideal radar image would integrate over the range and Doppler response functions. Figure 6.3 shows the simulated radar image obtained with simple cell averaging. The 10 GHz radar parameters were chosen to support an accurate approximation of $\sigma(\rho)$, aside from the coordinate mapping.

The potential degradation imposed by propagation effects depends on the radar bandwidth and the coherence time with respect to the coherent processing interval. With perfect frequency coherence, (6.25) represents the radar return after pulse compression. However, to generate the FPE solution it is necessary to consider the pulse-by-pulse mapping implicit in two-way field $\psi_\mathbf{k}^2(\rho + \mathbf{v_k}mT; \mathbf{R}_0, f_c)$. From the discussion in Chapter 4, the line represented by $\rho_0 + \mathbf{v_k}t$ is a scan across the measurement plane. To the extent that the structure is invariant over the coherent processing interval, ρ_0 values can be chosen to correspond to intervals separated by T. This is facilitated by converting $\psi_\mathbf{k}^2(\rho; \mathbf{R}_0, f_c)\tilde{\sigma}(\rho)$ to radar coordinates. If an accurate representation of sidelobes is not required, *pseudo-pulses* can be generated simply by taking the inverse Fourier transform of the mapped and distorted image along the delay variable.

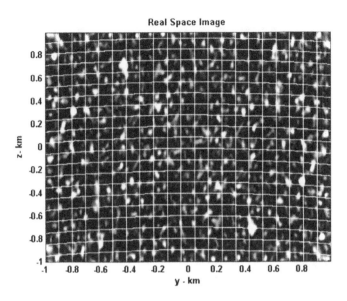

Figure 6.2   Realization of surface reflectivity with overlaid constant radar range (horizontal arcs) and constant range rate (vertical lines) contours.

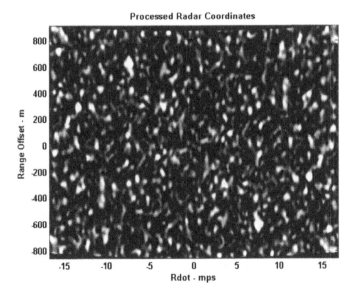

Figure 6.3   Processed radar image constructed by cell averaging the data shown in Figure 6.2.

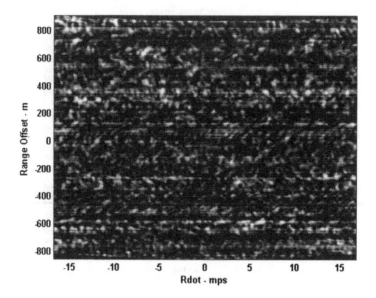

Figure 6.4   SAR image after two-way propagation through an intervening turbulent layer. The isotropic Kolmogorov turburence corresponded to a one-way SI index of 0.49.

For the 10 GHz aircraft model, clear-air turbulence is the most likely source of propagation structure. The disturbed radar image shown in Figure 6.4 was constructed by imposing the image plane structure for a disturbed layer at 12,000 ft scaled to induce a one-way disturbance with $SI = 0.49$ at the image. As one might suspect, the phase structure destroys the Doppler isolation of the scattering centers. Even so, the range resolution preserves the relative intensity of the structure in range. SAR images are routinely reconstructed to correct phase errors that result from the random physical motion of the aircraft. The same procedures reconstruct the degraded image shown here, up to the amplitude errors. However, pursuing this aspect of radar processing is beyond the scope of this book. The objective here is to demonstrate how to generate and implement realizations of propagation disturbances that can be incorporated into SAR and ISAR processing algorithms.

6.2 BOUNDARY SURFACES

To develop a formal procedure for evaluating propagation in a weakly inhomogeneous medium in the presence of an irregular boundary, it will be necessary to review the essential elements of boundary scattering theory. The strategy is to restructure the boundary integral equation (BIE) so that conditions

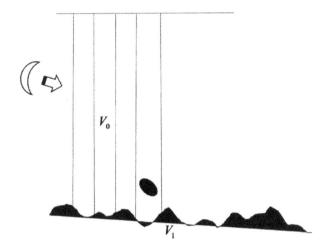

Figure 6.5   Schematic representation of forward propagation in the presence of a compact scattering object and a boundary surface.

under which backward propagation can be neglected are transparent. In the absence of backscatter interactions, boundary scattering theory can be aligned with the FPE and solved simultaneously at the same level of approximation. Although the procedure is tractable for some applications, the computational demands are significant. The example used to illustrate the methodology is two-dimensional, but it provides a framework to assess procedures that might be carried to a fully three-dimensional solution. It also provides a framework for assessing the differences between boundary integral methods and the more commonly used PWE-based methods.

6.2.1   Boundary Scattering Theory

A schematic diagram of a propagation environment that contains both an irregular boundary surface and a compact scattering object is shown in Figure 6.5. The volume V_0 is an exterior region that supports freely propagating waves, ultimately to be filled with weakly inhomogeneous structure. The volume V_1 is an interior region separated from V_0 by surfaces where the constitutive properties change discontinuously. A perfectly conducting boundary eliminates the penetration of fields from the exterior region into V_1. The scattering objects of primary interest are highly conducting and thereby penetrable only within a thin boundary layer. The penetration depth is referred to as skin depth. When it is small, imposing the boundary conditions is greatly simplified.

6.2.1.1 Boundary Integral Representation   The fields within the two volumes shown schematically in Figure 6.5 can be fully characterized by boundary integral representations of the electric field and the magnetic field.   Two equivalent boundary integral representations characterize the scattered and interior electric and magnetic fields.   The formal equivalence stems from the fact that the fields in the source-free interior and exterior regions are connected by Maxwell's equations.   For the development here, the boundary integral representation for the electric field will suffice.   The scattered field is defined as

$$\mathbf{E}_s(\mathbf{r}) = \mathbf{E}(\mathbf{r}) - \mathbf{E}_i(\mathbf{r}) \text{ in } V_0, \tag{6.28}$$

where $\mathbf{E}_i(\mathbf{r})$ is the impressed incident or source field.   The source field originates within V_0.   The development of the boundary integral equations that represent the exterior fields can be found in most graduate-level EM textbooks, for example, Chapter VI of *Electromagnetic Wave Theory* [10].   The electric field representation is written here as

$$\left\{ \begin{array}{l} \mathbf{E}_s(\mathbf{r}) \text{ for } \mathbf{r} \in V_0 \\ -\mathbf{E}_i(\mathbf{r}) \text{ for } \mathbf{r} \notin V_0 \end{array} \right\} = i\omega\mu_0 \iint \overline{\Gamma}_0(\mathbf{r},\mathbf{r}') \cdot [\mathbf{n}_0 \times \mathbf{H}(\mathbf{r}')] \, dS$$

$$+ \iint \nabla \times \overline{\Gamma}_0(\mathbf{r},\mathbf{r}') \cdot [\mathbf{n}_0 \times \mathbf{E}(\mathbf{r}')] \, dS, \tag{6.29}$$

where

$$\overline{\Gamma}(\mathbf{r},\mathbf{r}') = \left(\overline{\mathbf{I}} + \frac{1}{k^2} \nabla\nabla \right) G(\mathbf{r},\mathbf{r}'), \tag{6.30}$$

is the dyadic Green function, and $\mathbf{n}_0$ is a unit vector normal to the surface pointing into the exterior region V_0.   The scattered field, $\mathbf{E}_s(\mathbf{r})$, is supported by the surface currents $\mathbf{n}_0 \times \mathbf{E}(\mathbf{r}')$ and $\mathbf{n}_0 \times \mathbf{H}(\mathbf{r}')$, which are unknown.   Note that within V_1 the field supported by the surface currents cancels the incident field, which effectively isolates the interior region.

The corresponding boundary integral representation of the interior field, denoted $\mathbf{E}_t(\mathbf{r})$, is

$$\left\{ \begin{array}{l} \mathbf{E}_t(\mathbf{r}) \text{ for } \mathbf{r} \in V_1 \\ 0 \text{ for } \mathbf{r} \notin V_1 \end{array} \right\} = i\omega\mu_1 \iint \overline{\Gamma}_1(\mathbf{r},\mathbf{r}') \cdot [\mathbf{n}_0 \times \mathbf{H}(\mathbf{r}')] \, dS$$

$$+ \iint \nabla \times \overline{\Gamma}_1(\mathbf{r},\mathbf{r}') \cdot [\mathbf{n}_0 \times \mathbf{E}(\mathbf{r}')] \, dS, \tag{6.31}$$

where

$$\mathbf{E}_t(\mathbf{r}) = \mathbf{E}(\mathbf{r}) - \mathbf{E}_i(\mathbf{r}) \text{ in } V_1. \tag{6.32}$$

The dyadic Green function for the interior medium, as indicated by the 1 subscript, is evaluated with the interior constitutive parameters.   Note that the interior integral supports a null field in the exterior region.   The power of the boundary integral representations is that they represent the entire effect of the complementary region by induced sources on the boundary surface.   In effect, a three-dimensional problem is reduced to two dimensions.

6.2.1.2 Boundary Integral Equations   Because the surface currents in (6.29) and (6.31) are identical, the pair of boundary integral equations are sufficient to determine the unknown tangential fields $\mathbf{n}_0 \times \mathbf{E}(\mathbf{r}')$ and $\mathbf{n}_0 \times \mathbf{H}(\mathbf{r}')$. The system is overdetermined because the unknowns are confined to a surface. Solutions can be constructed in different ways, but the standard didactic approach first forms the cross products of the surface normal with the electric field and with the magnetic induction field.   Then the limit is taken as $\mathbf{r}$ approaches the surface along the normal direction. Because the dyadic Green function terms are singular on the surface, this limit must be evaluated carefully. Solving BIEs requires considerable expertise as even a cursory reading of Morita, *Integral Equation Methods for Electromagnetics* [120] will show.

The magnetic field integral equations (MFIEs) obtained by applying the procedure described above are

$$\frac{1}{2}\mathbf{a}(\mathbf{r}_s) = \mathbf{n}_0 \times \mathbf{H}_i(\mathbf{r}_s) - ik_0/\eta_0 \mathbf{n}_0 \times \iint \overline{\boldsymbol{\Gamma}}_0(\mathbf{r}_s, \mathbf{r}_s') \cdot \mathbf{b}(\mathbf{r}_s') dS$$

$$+\mathbf{n}_0 \times \iint \nabla \times \overline{\boldsymbol{\Gamma}}_0(\mathbf{r}_s, \mathbf{r}_s') \cdot \mathbf{a}(\mathbf{r}_s') \, dS \tag{6.33}$$

and

$$\frac{1}{2}\mathbf{a}(\mathbf{r}_s) = ik_1/\eta_1 \mathbf{n}_0 \times \iint \overline{\boldsymbol{\Gamma}}_1(\mathbf{r}_s, \mathbf{r}_s') \cdot \mathbf{b}(\mathbf{r}_s') \, dS$$

$$-\mathbf{n}_0 \times \iint \nabla \times \overline{\boldsymbol{\Gamma}}_1(\mathbf{r}_s, \mathbf{r}_s') \cdot \mathbf{a}(\mathbf{r}_s') \, dS, \tag{6.34}$$

where

$$\left\{ \begin{array}{c} \mathbf{a}(\mathbf{r}_s) \\ \mathbf{b}(\mathbf{r}_s) \end{array} \right\} = \left\{ \begin{array}{c} \mathbf{n}_0 \times \mathbf{H} \\ \mathbf{n}_0 \times \mathbf{E} \end{array} \right\}, \tag{6.35}$$

and the s subscript means "on the surface." There are complementary electric field integral equations (EFIEs) that can be used as well. The electric current $\mathbf{a}(\mathbf{r}_s)$ usually dominates the magnetic current $\mathbf{b}(\mathbf{r}_s)$, which is zero for perfectly conducting boundaries.   To solve the MFIE, (6.34) is manipulated first to obtain a non-local linear functional relation between $\mathbf{b}(\mathbf{r}_s)$ and $\mathbf{a}(\mathbf{r}_s)$. For a highly conducting boundary, the non-local integral relation can be replaced by a local matrix multiplication. This is referred to as an impedance boundary condition [121], formally

$$\mathbf{b}(\mathbf{r}_s) = Z(\mathbf{r})\mathbf{a}(\mathbf{r}_s), \tag{6.36}$$

where $Z(\mathbf{r})$ is a function representing the surface impedance. Substituting (6.36) for $\mathbf{b}(\mathbf{r}_s')$ in (6.33) leads to the following integral equation

$$\mathbf{a}(\mathbf{r}_s) + \boldsymbol{\Lambda}(\mathbf{r_s}; \mathbf{r}_s') \ominus \mathbf{a}(\mathbf{r}_s') = \frac{1}{2}\mathbf{n}_0 \times \mathbf{H}_i(\mathbf{r}_s), \tag{6.37}$$

where $\boldsymbol{\Lambda}(\mathbf{r_s}; \mathbf{r}_s') \ominus \mathbf{a}(\mathbf{r}_s')$ is defined as

$$\mathbf{\Lambda}(\mathbf{r}_s; \mathbf{r}'_s) \ominus \mathbf{a}(\mathbf{r}'_s) = \frac{1}{2}\mathbf{n}_0 \times \iint \left[\nabla \times \overline{\mathbf{\Gamma}}_0(\mathbf{r}_s, \mathbf{r}'_s) \cdot \mathbf{a}(\mathbf{r}'_s) \right.$$
$$\left. - ik_0/\eta_0 \overline{\mathbf{\Gamma}}_0(\mathbf{r}_s, \mathbf{r}'_s) \cdot (Z(\mathbf{r}'_s)\mathbf{a}(\mathbf{r}'_s)) \right] dS. \quad (6.38)$$

6.2.1.3 BIE Solutions Because there are two forms of the BIEs, which can be used independently or in combination, there is considerable latitude in the problem formulation stage. Nonetheless, solving BIEs is a formidable task that requires sub-wavelength sampling, and considerable numerical computational skill and EM insight. In any case, to translate the BIE into a form that in principle can be solved numerically, a surface mesh must be constructed that defines local triangular surface facets that uniquely define a normal vector. Each surface facet becomes a scattering center that can be represented as simply as a point weighting or as a summation of basis functions that represent the local field structure. Furthermore, because of the vector nature of the fields, each vector has three unknown components that must be accommodated. Nonetheless, the solution can be formulated as a system of linear equations of the form

$$\overline{\Lambda}\vec{a} = \vec{s}. \quad (6.39)$$

The square matrix $\overline{\Lambda}$ can be thought of as a super matrix with elements that are 3×3 matrices that operate on sub-vectors within the vector $\vec{a}$. The critical point is that the elements of $\overline{\Lambda}$ represent the interaction of a point on the surface, which is assigned a row index, with every other point on the surface, which are assigned column indices. The unknown vector $\vec{a}$ has elements that define the surface current at each facet. The vector $\vec{s}$ represents the known incident tangential field component at each facet on the surface. The fact that the elements of the matrices and vectors are themselves vectors is a data organization problem, albeit a critical one. The simpler vector notation is used here to emphasize the matrix form that can be exploited.

Once $\vec{a}$ is determined, a similar translation of (6.29) to matrix form is used to calculate the total field,

$$\mathbf{E}(\mathbf{r}) = \mathbf{E}_i(\mathbf{r}) + \overline{\Pi}(\mathbf{r}; \vec{\mathbf{r}}_s)\vec{a}. \quad (6.40)$$

Here $\overline{\Pi}(\mathbf{r}; \vec{\mathbf{r}}_s)$ is a matrix of vector functions that defines the field at any point in the propagation space as a summation over the discrete source vectors on the surface. Most remote sensing applications use approximations that do not reconcile the potential multiple interactions among the scattering centers. The Kirchhoff approximation, which is suggested by (6.37), substitutes $\frac{1}{2}\mathbf{n}_0 \times \mathbf{H}_i(\mathbf{r}_s)$ for $\mathbf{a}(\mathbf{r}_s)$. The Kirchhoff approximation is not restricted to small scattering amplitudes, but it has the same defect as the scintillation weak scatter approximation in that it neglects the field modification induced by the surface interaction. For example, shadowing is neglected. The weak surface scatter approximation, which restricts scattering amplitudes, cannot be derived directly from the spatial-domain form of the BIEs. Fourier domain transformations of the boundary integral representations are used. The lowest order approximation gives scattered fields proportional to the two-dimensional Fourier decomposition of the surface height variation evaluated at the Bragg wavenumber. A combined Kirchhoff and Bragg scatter two-scale theory [122], usually modified by geometric shadow corrections, is used for most remote sensing applications. A complete treatment of this aspect of surface scattering theory can be found in Chapter 2 of *Theory of Microwave Remote Sensing*, by Tsang, Kong, and Shin [123].

6.2.1.4 BIE Forward Approximation

To introduce the forward approximation for surface scatter, (6.40) is written in FPE coordinates as

$$\mathbf{E}(x, \zeta) = \Theta\left(x, x_0; \zeta\right) \mathbf{E}_i(x_0, \zeta) + \mathbf{\Pi}(x, \zeta; \vec{\mathbf{r}}_s) \vec{\mathbf{a}}. \tag{6.41}$$

The propagation operator $\Theta\left(x, x_0; \zeta\right)$ introduced in Chapter 2 is used to calculate the incident field in the plane x generated by the source field in the plane at $x = x_0$. In the standard application of boundary scattering theory, the sample mesh used to transform (6.37) into matrix form is constructed without consideration of the source field. A single mesh sampling is used for all source vectors $\vec{\mathbf{s}}$. However, the forward approximation is intimately tied to the propagation reference direction, which must be chosen so that propagation in the forward hemisphere totally dominates any propagating in the opposite hemisphere. Since scintillation phenomena are driven entirely by forward propagation, neglecting backscatter has no practical consequences. In remote sensing applications, that neglected backscatter may be the desired connection to the propagation media. The forward approximation only neglects backscatter in the media interaction computation. The possibility of using the FPE solution as a source function to estimate the backscatter has already been introduced (see Section 2.1.3). The idea of using this type of approximation in surface scattering theory is not new.

The method of ordered multiple interactions (MOMI) developed by Adams [124] is of particular interest in this regard. The MOMI method uses a decomposition of the matrix $\overline{\Lambda}$ into upper and lower triangular components to build a recursive series. The attractive feature of the MOMI recursion is that the first iteration often provides an adequate approximation to the complete solution. Unfortunately, when this is not the case the convergence of the MOMI recursion is poor. In any case, exploitation of the leading term in this type of recursion is the key to marrying the forward approximation and boundary effects into the FPE formalism.

In Rino, Doniger, and Martinez [125], the MOMI method was applied with *source-directed slice sampling*. That is, the surface samples are confined to planes normal to the reference propagation direction. To pursue this concept, consider a surface boundary defined at each point in the xy plane by the functional relation

$$z = f_s(x, y). \tag{6.42}$$

The sample vector

$$\vec{r}_s^n = [x_n, y_m, f_s(x_n, y_m)] \text{ for } 1 \leq m \leq M_n$$

identifies the surface samples in the plane at $x = x_n$. Now consider ordering the elements of $\vec{a}$ by slice planes. The first M_1 elements correspond to plane 1, the next M_2 elements to plane 2, and so on. The linear system to be solved can be organized in the block super matrix

$$
\begin{bmatrix}
\overline{\Lambda}_{11} & \overline{\Lambda}_{12} & \cdots & \overline{\Lambda}_{1N} \\
\overline{\Lambda}_{21} & \overline{\Lambda}_{22} & \cdots & \overline{\Lambda}_{2N} \\
\vdots & \vdots & \ddots & \vdots \\
\overline{\Lambda}_{N1} & \overline{\Lambda}_{N2} & \cdots & \overline{\Lambda}_{NN}
\end{bmatrix}
\begin{bmatrix}
\vec{a}_1 \\
\vec{a}_2 \\
\vdots \\
\vec{a}_N
\end{bmatrix}
=
\begin{bmatrix}
\vec{s}_1 \\
\vec{s}_2 \\
\vdots \\
\vec{s}_N
\end{bmatrix},
\tag{6.43}
$$

where $\vec{a}_n$ represents the current vector samples confined to the nth slice plane. The elements $\overline{\Lambda}_{nm}$ are $M_n \times M_m$ matrices of 3×3 sub-matrices that define the interactions between elements in the nth and the mth slice planes. The square matrices on the diagonal characterize the mutual interactions within the slice.

With slice sampling, the super matrix structure has a precise meaning. The sub-matrices above the diagonal ($\overline{\Lambda}_{nm}$ with $m > n$) characterize backward scattering interactions. The sub-matrices below the diagonal ($\overline{\Lambda}_{nm}$ with $m < n$) correspond to forward interactions. With the backward interactions set equal to zero, (6.43) can be solved by the recursion

$$\vec{a}_1 = \overline{\Lambda}_{11}^{-1} \vec{s}_1 \tag{6.44}$$

$$\vec{a}_n = \overline{\Lambda}_{nn}^{-1} \left[\vec{s}_n - \sum_{n'=1}^{n-1} \overline{\Lambda}_{nn'} \vec{a}_{n'} \right] \text{ for } 2 \leq n \leq N. \tag{6.45}$$

The solution evolves forward one plane at a time. In any slice plane the total field can be computed as

$$\mathbf{E}(x_n, \zeta) = \epsilon(\zeta) \left(\Theta(x_n, x_0; \zeta) \mathbf{E}_i(x_0, \zeta) + \sum_{n'=n_0}^{n-1} \Pi(x_n, \zeta; \vec{r}_s^{n'}) \vec{a}_{n'} \right), \tag{6.46}$$

where $n_0 = 1$, and

$$\epsilon(\zeta) = \begin{cases} 1 & \text{for } \zeta > \zeta_s \\ 1/2 & \text{for } \zeta = \zeta_s \end{cases}. \tag{6.47}$$

The 1/2 factor compensates for the singular behavior of the boundary integral at the surface. Although the solution is a forward recursion, it is not in a form that can be married to the split-step FPE solution.

For a split-step FPE computation, the propagation space must be divided into layers comprised of M BIE slices. Typically, M is greater than 100, reflecting the very different wavelength-dependent sampling requirements for the split-step FPE solution and the BIE solution. Aggregation is achieved by using the total field defined by (6.46) as the incident field to compute $\vec{s}_n$. This substitution is justified by the fact that the total field in a plane carries all the information needed to propagate the field forward in the absence of scattering interactions. However, because the source fields are singular at their point of origin, the scattered field contributions are not accurately represented near the singular points with a Fourier transformation. To see the problem explicitly, note that (2.7) shows that the scalar Green function can be computed with a two-dimensional Fourier transformation, but any fixed grid computation will rapidly deteriorate as the singularity is approached. To minimize this error, the redefined incident field should not be used in the current layer.

A forward marching recursion that maintains a buffer is implemented as follows. Let each layer start at slice $Ml + 1$ for $l = 0, 1, \cdots$. Assume that the total field at $n = Ml + 1$ is known and the source contributions $\vec{a}_n$ for $Ml + 1 \leq n \leq Ml + M$. The next set of $M - 1$ source functions are computed as follows:

$$\vec{a}_n = \overline{\Lambda}_{nn}^{-1} \vec{s}_n \qquad\qquad \text{for } n = Ml + 1 \quad (6.48)$$

$$\vec{a}_n = \overline{\Lambda}_{nn}^{-1} \left[\vec{s}_n - \sum_{n'=Ml-M}^{n-1} \overline{\Lambda}_{nn'} \vec{a}_{n'} \right] \begin{array}{l} \text{for } Ml + 2 \leq n \\ \leq Ml + M - 1 \end{array} . \quad (6.49)$$

The excitation field propagates at least one layer before contributing to $\vec{s}_n$. However, the new source functions are updated by all the source functions in the previous layer. Once the source computation is complete, the total field at $n = Ml$ is computed using (6.46) with $n_0 = Ml - M + 1$. Note that (6.46) has the FPE propagation step included. The split-step cycle is completed by incorporating the phase perturbation associated with the buffer layer. The split-step FPE BIE algorithm can be summarized as follows:

1. $\begin{cases} \text{Calculate } \vec{a}_n \text{ for } 1 \leq n \leq M \text{ using (6.45).} \\ \text{Save } \mathbf{E}_i(x_1, \zeta) \text{ as } \mathbf{E}(x_1, \zeta). \\ \text{Set } l = 0. \end{cases}$
2. Increment $l = l + 1$ and evaluate (6.48) and (6.49).
3. Compute $\mathbf{E}(x_{Ml}, \zeta)$ using (6.46) with $n_0 = Ml - M + 1$.
4. Impose phase perturbation on $\mathbf{E}(x_{Ml}, \zeta)$.
5. Repeat 2, 3, and 4 for each layer.

6.2.2 FPE Solution with Boundary Example

The example to be presented here uses the surface scatter formulation described in Rino and Ngo [126] and Rino and Kruger [26]. The complete two-dimensional BIE formulation and impedance boundary conditions are summarized in Appendix A.4 for reference. The results are useful for testing procedures prior to attacking fully three-dimensional problems. Two-dimensional surfaces introduce no polarization coupling. Thus, the cardinal polarizations can be evaluated independently. For a horizontally polarized incident wave, $H_y = E_x = E_z = 0$. For a vertically polarized wave, $E_y = H_x = H_z = 0$. To introduce notation consistent with the scalar field development to this point, let $\psi_h^+(x, z) = E_y$ for horizontal polarization and $\psi_v^+(x, z) = H_y$ for vertical polarization. For a perfectly conducting surface the simplest BIEs for horizontal and vertical polarization are, respectively,

$$
\frac{\partial \psi_h^+}{\partial N}(s) = 2\frac{\partial \psi_h^i}{\partial N}(s)
$$
$$
- \frac{i}{2} \int_{s_0}^s \frac{\partial H_0^{(1)}(k\Delta r(s, s'))}{\partial N} \frac{\partial \psi_h^+}{\partial N'}(s')ds' \qquad (6.50)
$$

and

$$
\psi_v^+(s) = 2\psi_v^i(s)
$$
$$
+ \frac{i}{2} \int_{s_0}^s \frac{\partial H_0^{(1)}(k\Delta r(s, s'))}{\partial N'} \psi_v^+(s)ds'. \qquad (6.51)
$$

The variable s is a continuous variable on the surface, and $\Delta r(s, s')$ represents the geometric distance between the two points on the surface at s and s'. A point on the surface has the vector location $\mathbf{r}_s = [x(s), z(s)]$, where $x(s)$ and $z(s)$ are functions specifying the coordinates of the surface point at s. If $z = f(x)$ is single valued, then $x(s) = x$. Otherwise, $x(s)$ identifies the single-valued branch. It follows that the normal derivative can be evaluated as

$$
\frac{\partial}{\partial N} = -\frac{\partial x(s)}{\partial s}\frac{\partial}{\partial x} + \frac{\partial z(s)}{\partial s}\frac{\partial}{\partial z}. \qquad (6.52)
$$

Once the BIEs have been solved, the horizontal and the vertical fields above the surface are computed respectively as

$$
\psi_h^+(x, z) = \psi_h^i(x, z) - \frac{i}{4}\int_{s_0}^s H_0^{(1)}(k\Delta r(x, z, s'))\frac{\partial \psi_h^+}{\partial N'}(s')ds' \qquad (6.53)
$$

and

$$
\psi_v^+(x, z) = \psi_v^i(x, z) + \frac{i}{4}\int_{s_0}^s \frac{\partial H_0^{(1)}(k\Delta r(x, z, s'))}{\partial N'}\frac{\partial \psi_h^+}{\partial N'}(s')ds'. \qquad (6.54)
$$

The simultaneous forward marching solution to the appropriate polarization-dependent BIE and the complementary forward propagation of the field using the algorithmic structure just described is straightforward. For completeness, discrete approximations to the BIE equations are summarized in Appendix

A.4. The propagation step can be evaluated as follows for a single-valued surface:

$$\psi(x, f(x)) = \int \widehat{\psi}(x, k_z) \exp\{ik_x(k_z)x\} \exp\{ik_z f(x)\} \frac{dk_z}{2\pi} \quad (6.55)$$

$$\frac{\partial \psi}{\partial N}(x, f(x)) = \int \widehat{\psi}(x, k_z)[-ik_x(k_z) + ik_z f'(x)]$$

$$\times \exp\{ik_x(k_z)x\} \exp\{ik_z f(x)\} \frac{dk_z}{2\pi}. \quad (6.56)$$

A single Fourier decomposition is used to propagate both the field and its normal derivative.

6.2.2.1 Surface Multipath

The first forward-propagation example with a boundary is an extension of the first example in Chapter 2. A 10 GHz beam source 100 m above the surface is used here. The CRPL standard atmosphere refractivity profile with no structure establishes a downward density gradient that bends the beam toward the surface. Earth curvature over the 18 km propagation distance is not negligible, but to simplify the display no curvature is included in the surface height definition. In effect, $z_s = f(x) = 0$. The height dimension, z, was sampled at 4 samples per wavelength with 8192 samples. The propagation medium was divided into 512 slabs, each comprising a propagation layer, with 513 subsamples per slab. The x dimension was sampled at 4 samples per wavelength. The quarter-wavelength horizontal sampling is coarse for BIE solutions, which often requires 10 or more samples per wavelength. The aperture was also phased to point the beam 2° downward, which increases the range of surface interaction. A radar configuration with this geometry might well be used for surface backscatter measurement. The result, scaled to represent the propagation factor defined in Section 1.2.4, is shown on a dB scale in Figure 6.6. The intensity pattern is well understood as the interference between the direct ray and the surface reflection from the unique point that satisfies the Fresnel reflection condition at the point of measurement. Although the result is a validation of the method, because refraction is incorporated there is no analytic solution to which the result can be compared.

6.2.2.2 Periodic Surface Multipath

The second example repeats the computation for a sinusoidal surface with a peak height of 3 meters and a spatial wavelength of 1 km. The result is shown in Figure 6.7. Because of the decrease in excitation intensity due to beam spreading, the scattering from each successive wave crest is diminished accordingly. More interesting, but harder to see in the gray scale figure, is the shadowing in the wave troughs. At the sample density used the shadow boundaries are not precise, but the computation cannot be reproduced with the Kirchhoff approximation. For testing purposes, the one-dimensional surface scatter problem must be solved

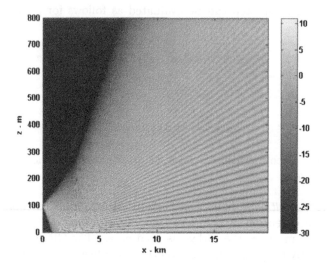

Figure 6.6   Forward propagation over a smooth conducting surface in a standard atmosphere.

numerically. Numerical simulations have been used extensively for surface scatter research [127]. One of the by-products of BIE solutions is the induced surface currents that support the scattering. Figure 6.8 shows the induced surface current for both smooth surface (upper frame) and the periodic surface (lower frame). The structure that can be seen in the period cycles shown in the lower frame of Figure 6.8 are mirrored in the scattered fields. The surface currents can be used to estimate backscatter levels as in Rino and Kruger [26], although small-scale structure must be present to support the backscatter.

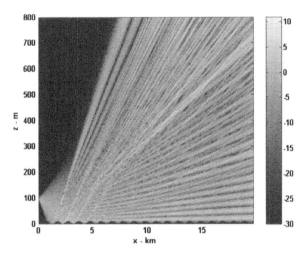

Figure 6.7 Forward propagation over a periodic conducting surface in a standard atmosphere.

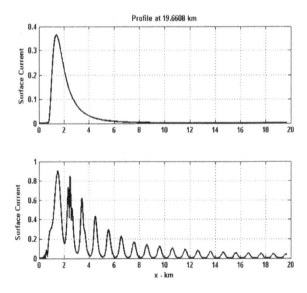

Figure 6.8 Comparison of induced boundary scatter source currents for smooth surface (upper frame) and periodic surface (lower frame).

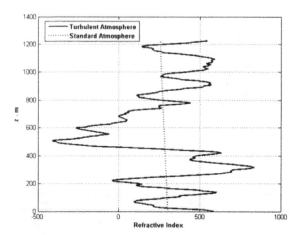

Figure 6.9 Realization of a turbulent atmosphere for propagation over a disturbed surface. The structure is imposed as a fractional modulation. The background refractivity is also shown (dashed).

6.2.2.3 Turbulent Layer Surface Multipath The final example introduces propagation through turbulence over the smooth reflecting layer. Figure 6.9 shows a realization of the turbulent structure from a single layer in refractivity units. The background atmosphere is superimposed (dashed) for reference. Because of the complexity of the multipath structure it is hard to pick out the additional modulation induced by the scintillation. Thus, Figure 6.10 shows a gray-scale plot of intensity difference in linear units between the result shown in Figure 6.6 and the corresponding result with turbulent structure. The results show the expected pattern of large scale vertical modulation, in this case a reduction at the higher altitudes and an enhancement at the lower altitudes. However, near the surface the pattern is more complex, indicating an interplay between the scattering and refraction. This is expected from field redefinition because the surface excitation field is influenced by the cumulative propagation effects that precede it. Figure 6.11 shows a comparison of the undisturbed and disturbed vertical profiles at the maximum propagation distance in the simulation.

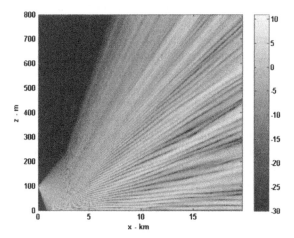

Figure 6.10 Intensity difference between propagation in standard atmosphere and standard atmosphere with and without turbulence.

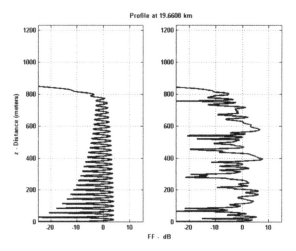

Figure 6.11 Comparison of vertical intensity profiles at maximum propagation distance. Left frame is standard atmosphere. Right frame is disturbed standard atmosphere.

6.2.3 Concluding Remarks

All of the two-dimensional examples shown here can be reproduced using PWE methods as detailed in the book by Levy [30]. A direct comparison between a PWE and FPE computation was presented in Rino and Kruger [26]. The results are similar enough that there is little motivation to implement the more complicated and computationally demanding BIE methods. However, the main reason for introducing the methodology here is to motivate the implementation of fully three-dimensional solutions. Considerable progress is being made with ray-tracing methods. The development has been stimulated by computer animation. However, with refraction and structure included, even ray tracing is too demanding for near-real-time execution. The ultimate solution may be a hybrid that uses ray tracing combined with a local diffraction computation.

APPENDIX A

A.1 FAR-FIELD APPROXIMATION

Introductory electromagnetics textbooks usually develop the far-field approximation by manipulating the spatial-domain integral representations of the fields. However, it is more in keeping with the development in this book to use the stationary phase approximation. Recall that the summation of vector plane waves represented by the two-dimensional Fourier construction,

$$\mathbf{F}(\varsigma) = \iint \widehat{\mathbf{F}}(\boldsymbol{\kappa}) \exp\{i\,(kg(\kappa)x + \boldsymbol{\kappa} \cdot \boldsymbol{\xi})\}\frac{d\boldsymbol{\kappa}}{(2\pi)^2}, \qquad (A.1)$$

satisfies the vector wave equation in an unbounded homogeneous medium. That is,

$$\nabla^2\mathbf{F} + k^2\mathbf{F} = 0. \qquad (A.2)$$

For reference,

$$\varsigma = x\widehat{\mathbf{a}}_x + \boldsymbol{\xi}, \qquad (A.3)$$

where x is the reference axis for propagation. The corresponding spatial wavenumber variables are

$$\mathbf{k} = [kg(\kappa), \boldsymbol{\kappa}], \qquad (A.4)$$

where

$$g(\kappa) = \sqrt{1 - (\kappa/k)^2}. \tag{A.5}$$

The argument of the exponent in (A.1) is $\mathbf{k} \cdot \boldsymbol{\varsigma}$.

For most applications x greatly exceeds the range of $\boldsymbol{\xi}$ in the measurement plane. Indeed, one often characterizes the field in terms of angles that translate to a fixed value of $\boldsymbol{\xi}/x$. Under these conditions, the method of stationary phase gives an approximation of the form

$$\mathbf{F}(\varsigma) \approx \frac{-i\widehat{\mathbf{F}}(\boldsymbol{\kappa}_s)}{2\pi\sqrt{|\Theta|}} \exp\{i\Theta(\kappa_y^s, \kappa_z^s)\}, \tag{A.6}$$

where

$$\Theta(\kappa_y, \kappa_z) = \kappa_y \xi_y + \kappa_z \xi_z + kg(\kappa)x \tag{A.7}$$

is the stationary point of the argument, and

$$|\Theta| = \frac{\partial^2 \Theta(\kappa_y, \kappa_z)}{\partial \kappa_y^2} \frac{\partial^2 \Theta(\kappa_y, \kappa_z)}{\partial \kappa_z^2} - \left(\frac{\partial^2 \Theta(\kappa_y, \kappa_z)}{\partial \kappa_y \partial \kappa_z}\right)^2 \Bigg|_{\kappa_y^s, \kappa_z^s}. \tag{A.8}$$

The s superscript refers to the stationary phase points where the derivatives of $\Phi(\kappa_y, \kappa_z)$ with respect to κ_y and κ_z are zero. By direct computation

$$\frac{\partial \Theta(\kappa_y, \kappa_z)}{\partial \kappa_{y,z}} = \xi_{y,z} - \frac{x K_{y,z}}{g(\kappa)} = 0 \tag{A.9}$$

determines the stationary point as follows:

$$g(\kappa_s) = 1/\sqrt{1 + (\xi/x)^2}, \tag{A.10}$$

where the ratio ξ/x is fixed, and

$$\kappa_{y,z}^s = \frac{k\xi_{yz}}{x\sqrt{1 + (\xi/x)^2}}. \tag{A.11}$$

Similarly,

$$\frac{\partial^2 \Theta(\kappa_y, \kappa_z)}{\partial \kappa_{y,z}^2} = -\frac{x}{kg(\kappa)}\left(1 + \frac{\kappa_{y,z}^2}{\kappa^2 g^2(\kappa)}\right), \tag{A.12}$$

and

$$\frac{\partial^2 \Theta(\kappa_y, \kappa_z)}{\partial \kappa_y \partial \kappa_x} = -\frac{x}{kg(\kappa)}\left(\frac{\kappa_y \kappa_z}{\kappa^2 g^2(\kappa)}\right). \tag{A.13}$$

Thus,

$$\sqrt{|\Theta|} = \frac{x}{kg(\kappa_s)} = \frac{r}{\kappa}, \tag{A.14}$$

and

$$\Phi(\kappa_y^s, \kappa_z^s) = kr. \tag{A.15}$$

Collecting the results,

$$\mathbf{F}(\varsigma) \approx \frac{-ikg(\kappa_s)\widehat{\mathbf{F}}(\kappa_s)}{2\pi} \frac{\exp\{ikr\}}{r}, \qquad (A.16)$$

which is a locally spherical wavefront modulated by the two-dimensional Fourier transform of the field in the reference plane.

A.2 BACKSCATTER

In Chapter 2 it was shown that (2.8) characterizes the scattered field from a structure region bounded by two planes at $x = 0$ and $x = L$. The integral representation applies if the modification of the incident field by the interaction is negligible.   Backscatter radars and ionospheric sounders are designed to detect scatter from disturbed regions in the atmosphere and in the ionosphere.   One usually assumes that the excitation field is the single plane wave $\mathbf{E}_o \exp\{i\mathbf{k}_i \cdot \mathbf{r}\}$. The correspondig two-dimensional spatial Fourier transform in the $x = 0$ plane is

$$\widehat{\mathbf{E}}(0; \boldsymbol{\kappa}') = \mathbf{E}_o \delta(\boldsymbol{\kappa}' - \boldsymbol{\kappa}_i)(2\pi)^2. \qquad (A.17)$$

With this substitution, (2.8) simplifies to

$$\begin{aligned}
\mathbf{E}_s(\mathbf{r}) &= i\mathbf{E}_o \iint \widehat{S}(\Delta\kappa^{\pm}(\boldsymbol{\kappa}, \boldsymbol{\kappa}_i)) \\
&\quad \times \frac{\exp\{-ikg(\kappa)x\}}{g(\kappa)} \exp\{i\boldsymbol{\kappa}\cdot\boldsymbol{\zeta}\}d\boldsymbol{\kappa}/(2\pi)^2.
\end{aligned} \qquad (A.18)$$

Upon comparing (A.1) with (A.18) it follows that in the far field

$$\mathbf{E}_s(\mathbf{r}) \approx \frac{-ikg(\kappa_i)\widehat{S}(\Delta\kappa^-(\boldsymbol{\kappa}_r, \boldsymbol{\kappa}_i))}{2\pi g(\kappa_r)} \widehat{\mathbf{E}}(0; \boldsymbol{\kappa}_i) \frac{\exp\{ikr\}}{r}. \qquad (A.19)$$

If $\mathbf{r}$ is in the strict backscatter direction $\boldsymbol{\kappa}_r = \boldsymbol{\kappa}_i$,   then $\Delta\kappa^-(\boldsymbol{\kappa}_r, \boldsymbol{\kappa}_i) = (2kg(\kappa_i), 0)$ and the $g(\kappa)$ factors cancel.

If the structure can be modeled as a homogeneous random process within the scattering volume it follows from (A.19) that the average backscattered power is proportional to the three-dimensional spectral density function

$$\left\langle \left| \widehat{S}(\Delta\kappa^-(\boldsymbol{\kappa}_r, \boldsymbol{\kappa}_i)) \right|^2 \right\rangle$$

evaluated at the Bragg wavenumber $\Delta\kappa^-(\boldsymbol{\kappa}_r, \boldsymbol{\kappa}_i)$. Because a radar measures delay, the range from where the backscatter originates can be determined from the mean speed of light in the medium. To the extent that the incident field is known, (A.19) can be inverted to estimate the average scattered intensity at the Bragg wavenumber. The illuminated volume is defined by the radar range resolution, the beam width, and the propagation direction.

A.3 ANISOTROPY TRANSFORMATIONS

A statistically homogeneous random field admits an autocorrelation function (ACF) that depends only on the position difference variable $\Delta\varsigma$. If the system is isotropic, the dependence varies only with the magnitude of the difference vector $y = \sqrt{|\Delta\varsigma|^2}$. Assuming that all structure scales have the same elongation, Singleton introduced a series of rotations and a scaling to transform an isotropic ACF to a general anisotropic form. Consider a vector in a reference topocentric coordinate system with the x axis pointing downward, the y axis pointing eastward, and the z axis pointing southward.   The rotations and scaling that affect the transformation can be written as

$$y = \sqrt{\left(\overline{D}_{ab}^{-1}\overline{U}_{\gamma_B}\overline{U}_{\psi_B}\overline{U}_{\phi_B}\Delta\varsigma\right)^T \left(\overline{D}_{ab}^{-1}\overline{U}_{\gamma_B}\overline{U}_{\psi_B}\overline{U}_{\phi_B}\Delta\varsigma\right)}, \qquad (A.20)$$

where $\overline{U}_{\phi_B}$, $\overline{U}_{\psi_B}$, and $\overline{U}_{\gamma_B}$ are rotation matrices that will be defined below, and

$$\overline{D}_{ab} = \begin{bmatrix} 1 & 0 & 0 \\ 0 & a & 0 \\ 0 & 0 & b \end{bmatrix}. \qquad (A.21)$$

The rotation

$$\overline{U}_{\phi_B} = \begin{bmatrix} 1 & 0 & 0 \\ 0 & \cos\phi_B & \sin\phi_B \\ 0 & -\sin\phi_B & \cos\phi_B \end{bmatrix} \qquad (A.22)$$

about the x axis aligns the meridian plane with the plane of the magnetic field in a new $x'y'z'$ system. The rotation

$$\overline{U}_{\psi_B} = \begin{bmatrix} \cos\psi_B & \sin\psi_B & 0 \\ -\sin\psi_B & \cos\psi_B & 0 \\ 0 & 0 & 1 \end{bmatrix} \qquad (A.23)$$

about z' through the dip angle ψ_B aligns the z'' axis with the principal irregularity axis.   A final rotation about y'' through an angle γ_B aligns the y''' axis with a second elongation axis

$$\overline{U}_{\gamma_B} = \begin{bmatrix} \cos\gamma_B & 0 & \sin\gamma_B \\ 0 & 1 & 0 \\ -\sin\gamma_B & 0 & \cos\gamma_B \end{bmatrix}. \qquad (A.24)$$

The combined rotations define the unitary transformation

$$\overline{U}_{\gamma_B}\overline{U}_{\psi_B}\overline{U}_{\phi_B} = \begin{bmatrix} c11 & c12 & c12 \\ c21 & c22 & c23 \\ c31 & c32 & c33 \end{bmatrix}, \qquad (A.25)$$

where

$$c11 = \cos\gamma_B \cos\psi_B \tag{A.26}$$

$$c12 = \cos\gamma_B \sin\psi_B \cos\phi_B - \sin\gamma_B \sin\phi_B \tag{A.27}$$

$$c13 = \cos\gamma_B \sin\psi_B \sin\phi_B + \sin\gamma_B \cos\phi_B \tag{A.28}$$

$$c21 = -\sin\psi_B \tag{A.29}$$

$$c22 = \cos\psi_B \cos\phi_B \tag{A.30}$$

$$c23 = \cos\psi_B \sin\phi_B \tag{A.31}$$

$$c31 = -\sin\gamma_B \cos\psi_B \tag{A.32}$$

$$c32 = -\sin\gamma_B \sin\psi_B \cos\phi_B - \cos\gamma_B \sin\phi_B \tag{A.33}$$

$$c33 = -\sin\gamma_B \sin\psi_B \sin\phi_B + \cos\gamma_B \cos\phi_B \tag{A.34}$$

The complete Singleton transformation can be written in vector matrix form as

$$\begin{bmatrix} r \\ s \\ t \end{bmatrix} = \overline{D}_{ab}^{-1} \overline{U}_{\gamma_B} \overline{U}_{\psi_B} \overline{U}_{\phi_B} \begin{bmatrix} x \\ y \\ z \end{bmatrix}. \tag{A.35}$$

Defining the product matrix explicitly

$$D_{ab}^{-1} U_{\gamma_B} U_{\psi_B} U_{\phi_B} = \begin{bmatrix} c11 & c12 & c13 \\ c21/a & c22/a & c23/a \\ c31/b & c32/b & c33/b \end{bmatrix} \tag{A.36}$$

It is convenient to summarize the results in terms of a single matrix $\overline{C}$ and the complementary matrix $\widehat{C}$ that defines constant spectral density function (SDF) intensity surfaces. The quadratic-function relation defined in the spatial domain is

$$y^2 = \Delta\varsigma^T \overline{C} \Delta\varsigma. \tag{A.37}$$

It is readily shown that the complementary transformations that impose the Singleton anisotropy in the spatial Fourier domain can be written as

$$q^2 = \kappa^T \widehat{C} \kappa, \tag{A.38}$$

where the elements of $\widehat{C}$ are obtained by replacing the occurrences in $\overline{C}$ of $1/a$ with a and the occurrences of $1/b$ with b. The quadratic form that defines the contours of constant correlation can be constructed as follows:

$$
\begin{aligned}
r^2 &= c11^2 x^2 + c12^2 y^2 + c13^2 z^2 \\
&\quad + 2\,(c11)\,(c12)\,xy + 2\,(c11)\,(c13)\,xz + (c12)\,(c13)\,yz \quad \text{(A.39)} \\
s^2 &= (c21/a)^2\,x^2 + (c22/a)^2\,y^2 + (c23/a)^2\,z^2 \\
&\quad + 2\,((c21)\,(c22)\,/a^2)\,xy + 2\,((c21)\,(c23)\,/a^2)\,xz \\
&\quad + ((c22)\,(c23)\,/a^2)\,yz \quad \text{(A.40)} \\
t^2 &= (c31/b)^2\,x^2 + (c32/b)^2\,y^2 + (c33/b)^2\,z^2 \\
&\quad + 2\,(c31)\,(c32)\,/b^2 xy + 2\,(c31)\,(c33)\,/b^2 xz \\
&\quad + (c32)\,(c33)\,/b^2 yz \quad \text{(A.41)}
\end{aligned}
$$

$$
\begin{aligned}
r^2 + s^2 + t^2 &= \left(c11^2 + (c21/a)^2 + (c31/b)^2\right) x^2 \\
&\quad + \left(c12^2 + (c22/a)^2 + (c32/b)^2\right) y^2 \\
&\quad + \left(c13^2 + (c23/a)^2 + (c33/b)^2\right) z^2 \\
&\quad + 2\,((c11)\,(c12) + (c21)\,(c22)\,/a^2 \\
&\quad + (c31)\,(c32)\,/b^2 xy) \\
&\quad + 2\,((c11)\,(c13) + (c21)\,(c23)\,/a^2 \\
&\quad + (c31)\,(c33)\,/b^2 xz) \\
&\quad + 2\,((c12)\,(c13) + (c22)\,(c23)\,/a^2 \\
&\quad + (c32)\,(c33)\,/b^2 yz)\,. \quad \text{(A.42)}
\end{aligned}
$$

In matrix form,

$$
y^2 = \Delta\boldsymbol{\xi}^T \overline{C} \Delta\boldsymbol{\xi}, \quad \text{(A.43)}
$$

where

$$
\begin{aligned}
C11 &= c11^2 + (c21/a)^2 + (c31/b)^2 & \text{(A.44)} \\
C22 &= c12^2 + (c22/a)^2 + (c32/b)^2 & \text{(A.45)} \\
C33 &= c13^2 + (c23/a)^2 + (c33/b)^2 & \text{(A.46)} \\
C12 &= C21 = (c11)\,(c12) + (c21)\,(c22)\,/a^2 + (c31)\,(c32)\,/b^2 & \text{(A.47)} \\
C13 &= C31 = (c11)\,(c13) + (c21)\,(c23)\,/a^2 + (c31)\,(c33)\,/b^2 & \text{(A.48)} \\
C23 &= C32 = (c12)\,(c13) + (c22)\,(c23)\,/a^2 + (c32)\,(c33)\,/b^2 & \text{(A.49)}
\end{aligned}
$$

The transformation that accommodates oblique propagation is implemented in the spatial wavenumber domain as

$$
\kappa_x = \tan\theta\,(\kappa_y \cos\phi + \kappa_z \sin\phi)\,, \quad \text{(A.50)}
$$

where θ and ϕ are the polar angles that define the principal propagation direction. It follows by direct computation that

$$
\begin{aligned}
q^2 &= \widehat{C}11 \tan^2 \theta \left(\kappa_y^2 \cos^2 \phi + 2\kappa_y \kappa_z \sin \phi \cos \phi + \kappa_z^2 \sin^2 \phi \right) + \widehat{C}22 \kappa_y^2 \\
&\quad + \widehat{C}33 \kappa_z^2 + 2\widehat{C}12 \tan \theta \left(\kappa_y \cos \phi + \kappa_z \sin \phi \right) \kappa_y \\
&\quad + 2\widehat{C}13 \tan \theta \left(\kappa_y \cos \phi + \kappa_z \sin \phi \right) \kappa_z + 2\widehat{C}23 \kappa_y \kappa_z
\end{aligned}
\tag{A.51}
$$

$$
\begin{aligned}
&= \kappa_y^2 \left(\widehat{C}11 \tan^2 \theta \cos^2 \phi + \widehat{C}22 + 2\widehat{C}12 \tan \theta \cos \phi \right) \\
&\quad + 2\kappa_y \kappa_z \left(\widehat{C}11 \tan^2 \theta \sin \phi \cos \phi + \widehat{C}23 \right. \\
&\quad \left. + \tan \theta \left(\widehat{C}12 \sin \phi + \widehat{C}13 \cos \phi \right) \right)
\end{aligned}
\tag{A.52}
$$

$$
+ \kappa_z^2 \left(\widehat{C}11 \tan^2 \theta \sin^2 \phi + \widehat{C}33 + 2\widehat{C}13 \tan \theta \sin \phi \right),
\tag{A.53}
$$

which is now two-dimensional. The two-dimensional quadratic form can be written as

$$
q^2 = A\kappa_y^2 + B\kappa_y \kappa_z + C\kappa_z^2,
\tag{A.54}
$$

where

$$
A = \widehat{C}22 + \widehat{C}11 \tan^2 \theta \cos^2 \phi + 2\widehat{C}12 \tan \theta \cos \phi
\tag{A.55}
$$

$$
\begin{aligned}
B &= 2 \left(\widehat{C}23 + \widehat{C}11 \tan^2 \theta \sin \phi \cos \phi \right. \\
&\quad \left. + \tan \theta \left(\widehat{C}12 \sin \phi + \widehat{C}13 \cos \phi \right) \right)
\end{aligned}
\tag{A.56}
$$

$$
C = \widehat{C}33 + \widehat{C}11 \tan^2 \theta \sin^2 \phi + 2\widehat{C}13 \tan \theta \sin \phi
\tag{A.57}
$$

The rotation that removes the cross term is used in the development. For completeness let

$$
\kappa_y = \kappa_y' \cos \vartheta + \kappa_z' \sin \vartheta
\tag{A.58}
$$

$$
\kappa_z = -\kappa_y' \sin \vartheta + \kappa_z' \cos \vartheta,
\tag{A.59}
$$

It follows that

$$
\begin{aligned}
q^2 &= A \left(\kappa_y'^2 \cos^2 \vartheta + 2\kappa_y' \kappa_z' \sin \vartheta \cos \vartheta + \kappa_z'^2 \sin^2 \vartheta \right) \\
&\quad + \left(\kappa_z'^2 - \kappa_y'^2 \right) B \cos \vartheta \sin \vartheta + \kappa_y' \kappa_z' B \cos 2\vartheta \\
&\quad + C \left(\kappa_y'^2 \sin^2 \vartheta - 2\kappa_y' \kappa_z' \sin \vartheta \cos \vartheta + \kappa_z'^2 \cos^2 \vartheta \right) \\
&= \kappa_y'^2 \left(A \cos^2 \vartheta - B \cos \vartheta \sin \vartheta + C \sin^2 \vartheta \right) \\
&\quad + 2\kappa_y' \kappa_z' \left((A - C) \sin \vartheta \cos \vartheta + B/2 \cos 2\vartheta \right) \\
&\quad + \kappa_z'^2 \left(A \sin^2 \vartheta + B \cos \vartheta \sin \vartheta + C \cos^2 \vartheta \right).
\end{aligned}
\tag{A.60}
$$

The product term is eliminated with

$$
\tan 2\vartheta = \frac{B/2}{(C - A)}.
\tag{A.61}
$$

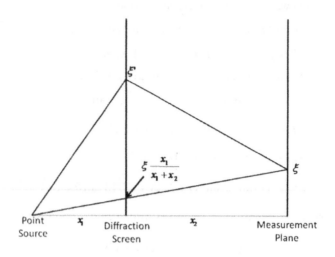

Figure A.1   Displaced and direct ray paths from point source to measurement plane.

In terms of A, B, and C,

$$\sin^2 \vartheta = 1/2 - (C - A)/(2D) \tag{A.62}$$

$$\cos^2 \vartheta = 1/2 + (C - A)/(2D) \tag{A.63}$$

$$\sin \vartheta \cos \vartheta = \frac{B/4}{D}, \tag{A.64}$$

where

$$D = \sqrt{(C - A)^2 + (B/2)^2}. \tag{A.65}$$

Finally, in the rotated coordinate system,

$$\begin{aligned} q^2 &= \kappa_y'^2 (A + C - D)/2 \\ &\quad + \kappa_z'^2 (A + C + D)/2. \end{aligned} \tag{A.66}$$

A.4   WAVEFRONT CURVATURE CORRECTION

Following a simple argument produced in a seminal paper by Ratcliffe [72], consider the path difference between the direct path shown in Figure A.1 and the displaced path in the screen given by

$$
\begin{aligned}
\Delta p &= \sqrt{x_1^2 + \xi'^2} + \sqrt{x_2^2 + (\xi' - \xi)^2} - \sqrt{(x_1 + x_2)^2 + \xi^2} \\
&\simeq \frac{1}{2x_2}\left(\left(\frac{x_1 + x_2}{x_1}\right)\xi'^2 - 2\xi'\xi - \left(\frac{x_1}{x_1 + x_2}\right)\xi^2\right) \\
&= \frac{1}{2x_2 CF}\left(\xi' - CF\xi\right)^2,
\end{aligned} \tag{A.67}
$$

where

$$
CF = \frac{x_1}{x_1 + x_2}. \tag{A.68}
$$

The measure Δp appears in most diffraction computations. The computation shows that scaling each spatial dimension by CF incorporates the effects of a point source in a plane wave computation. For oblique propagation, distances x_1 and x_2 refer to the distances along and transverse to the principal propagation direction, respectively. The generalization to a three-dimensional computation follows directly.

A.5 TWO-DIMENSIONAL BOUNDARY INTEGRALS

The boundary integral equation (BIE) field representations can be simplified considerably when the variation along one principal axis, for example y, can be neglected. In this case one integration in each of the BIE surface integrals can be evaluated explicitly. The resulting BIEs are one-dimensional with cylindrical-wave Hankel functions replacing the spherical-wave Green functions. With the time-variation convention $\exp\{-i\omega t\}$, Hankel functions of the first kind represent outgoing cylindrical waves that are proportional to $\exp(ik\rho)/\sqrt{\rho}$. All variation takes place in the xz plane and there is no polarization coupling. Let $\mathbf{E}$ and $\mathbf{H}$ represent the electric and magnetic field vectors with $\mathbf{n}_0$ the outward normal vector on the boundary surface and $\mathbf{u}$ a unit vector such that $\mathbf{u}_k = \mathbf{k}/k$. The wavenumber is k and the surface impedance is $\eta = \mu/\epsilon$. For horizontal polarization,

$$
\mathbf{n}_0 \times \mathbf{H} = \frac{i}{\eta k}\frac{\partial E_y}{\partial n_0}\mathbf{u}_y \tag{A.69}
$$

$$
\mathbf{n}_0 \times \mathbf{E} = E_y(\zeta)\mathbf{n}_0 \times \mathbf{u}_y. \tag{A.70}
$$

For vertical polarization,

$$
\mathbf{n}_0 \times \mathbf{E} = -\frac{i\eta}{k}\frac{\partial H_y}{\partial n_0}\mathbf{u}_y \tag{A.71}
$$

$$
\mathbf{n}_0 \times \mathbf{H} = H_y(\zeta)\mathbf{n}_0 \times \mathbf{u}_y. \tag{A.72}
$$

The independent variable ζ denotes a coordinate that admits a one-to-one mapping onto the surface contour. The two-dimensional BIEs summarized below are written in terms of y-directed scalar field components and their normal derivatives [120, Chapter 2.3].

A.5.0.1 Horizontal Polarization (Transverse Electric) For a horizontally polarized incident wave $H_y = E_x = E_z = 0$. The electric field integral equations (EFIEs) for the exterior and interior fields are given by

$$\frac{1}{2}E_y(\zeta) = E_{iy}(\zeta) - \int \frac{i}{4}H_0^{(1)}(k_0\Delta\rho_s)\frac{\partial E_y}{\partial n_0}(\zeta')\,d\zeta'$$
$$+ \int \frac{i}{4}\frac{\partial H_0^{(1)}}{\partial n_0'}(k_0\Delta\rho_s)E_y(\zeta')\,d\zeta', \tag{A.73}$$

and

$$\frac{1}{2}E_y(\zeta) = \left(\frac{\eta_1}{k_0\eta_0}\right)\int \frac{ik_1}{4}H_0^{(1)}(k_1\Delta\rho_s)\frac{\partial E_y}{\partial n_0}(\zeta')\,d\zeta'$$
$$- \int \frac{i}{4}\frac{\partial H_0^{(1)}}{\partial n_0'}(k_1\Delta\rho_s)E_y(\zeta')\,d\zeta', \tag{A.74}$$

where $\Delta\rho = \sqrt{(x-x')^2 + (z-f(x'))^2}$, and $\Delta\rho_s$ means z has been replaced by $f(x)$. Thus, $\Delta\rho_s$ is implicitly a function of ζ and ζ'. The corresponding magnetic field integral equations (MFIEs) for the exterior and interior fields are

$$\frac{1}{2}\frac{\partial E_y}{\partial n_0}(\zeta) = \frac{\partial E_{iy}}{\partial n_0}(\zeta)$$
$$+ \int \frac{ik_0^2}{4}H_0^{(1)}(k_0\Delta\rho_s)\left(1 + \frac{\partial f}{\partial x}\frac{\partial f}{\partial x'}\right)E_y(\zeta')\,d\zeta'$$
$$+ \int \frac{i}{4}\Omega(k_0,\zeta,\zeta')\,E_y(\zeta')\,d\zeta'$$
$$- \int \frac{i}{4}\frac{\partial H_0^{(1)}}{\partial n_0}(k_0\Delta\rho_s)\frac{\partial E_y}{\partial n_0}(\zeta')\,d\zeta', \tag{A.75}$$

and

$$\frac{1}{2}\frac{\partial E_y}{\partial n_0}(\zeta) = -\left(\frac{k_0\eta_0}{\eta_1}\right)\int \frac{ik_1}{4}H_0^{(1)}(k_1\Delta\rho_s)$$
$$\times\left(1 + \frac{\partial f}{\partial x}\frac{\partial f}{\partial x'}\right)E_y(\zeta')\,d\zeta'$$
$$- \left(\frac{k_0\eta_0}{\eta_1}\right)\int \frac{i}{4k_1}\Omega(k_1,\zeta,\zeta')$$
$$\times E_y(\zeta')\,d\zeta'$$
$$+ \int \frac{i}{4}\frac{\partial H_0^{(1)}}{\partial n_0}(k_1\Delta\rho_s)\frac{\partial E_y}{\partial n_0'}(\zeta')\,d\zeta', \tag{A.76}$$

where

$$\Omega(k_0, \zeta, \zeta') = \frac{\partial f}{\partial x}\left(\frac{\partial^2 H_0^{(1)}}{\partial z \partial x} + \frac{\partial^2 H_0^{(1)}}{\partial z^2}\frac{\partial f}{\partial x'}\right)$$
$$+ \left(\frac{\partial^2 H_0^{(1)}}{\partial x^2} + \frac{\partial^2 H_0^{(1)}}{\partial x \partial z}\frac{\partial f}{\partial x'}\right). \tag{A.77}$$

The MFIE equations cannot be obtained simply by taking the normal derivative of the EFIE equations unless $E_y(\zeta) = 0$.

A.5.0.2 Vertical Polarization (Transverse Magnetic) For a vertically polarized incident wave $E_y = H_x = H_z = 0$. The EFIEs for the interior and exterior field are given by

$$\frac{1}{2}\frac{\partial H_y}{\partial n_0}(\zeta) = \frac{\partial H_{iy}}{\partial n_0}(\zeta) +$$
$$\int \frac{ik_0^2}{4} H_0^{(1)}(k_0\Delta\rho_s)\left(1 + \frac{\partial f}{\partial x}\frac{\partial f}{\partial x'}\right) H_y(\zeta')\,d\zeta'$$
$$+ \int \frac{i}{4}\Omega(k_0, \zeta, \zeta')H_y(\zeta')d\zeta'$$
$$- \int \frac{i}{4}\frac{\partial H_0^{(1)}}{\partial n_0}(k_0\Delta\rho_s)\frac{\partial H_y}{\partial n_0'}\,d\zeta', \tag{A.78}$$

and

$$\frac{1}{2}\frac{\partial H_y}{\partial n_0}(\zeta) = -\left(\frac{k_0\eta_1}{\eta_0}\right)\int \frac{ik_1}{4}H_0^{(1)}(k_1\Delta\rho_s)$$
$$\times \left(1 + \frac{\partial f}{\partial x}\frac{\partial f}{\partial x'}\right)H_y(\zeta')\,d\zeta'$$
$$- \left(\frac{k_0\eta_1}{\eta_0}\right)\int \frac{i}{4k_1}\Omega(k_1, \zeta, \zeta')H_y(\zeta')d\zeta'$$
$$+ \int \frac{i}{4}\frac{\partial H_0^{(1)}}{\partial n_0}(k_1\Delta\rho_s)\frac{\partial H_y}{\partial n_0'}(\zeta')\,d\zeta'. \tag{A.79}$$

The MFIEs reduce to

$$\frac{1}{2}H_y(\zeta) = H_{iy}(\zeta) - \int \frac{i}{4}H_0^{(1)}(k_0\Delta\rho_s)\frac{\partial H_y}{\partial n_0}(\zeta')\,d\zeta'$$
$$+ \int \frac{i}{4}\frac{\partial H_0^{(1)}}{\partial n_0'}(k_0\Delta\rho_s)H_y(\zeta')\,d\zeta', \tag{A.80}$$

and

$$\frac{1}{2}H_y\left(\zeta\right) = \left(\frac{\eta_0}{k_0\eta_1}\right)\int \frac{ik_1}{4}H_0^{(1)}(k_1\Delta\rho_s)\frac{\partial H_y}{\partial n_0}\left(\zeta'\right)d\zeta'$$
$$-\int \frac{i}{4}\frac{\partial H_0^{(1)}}{\partial n_0'}(k_1\Delta\rho_s)H_y\left(\zeta'\right)d\zeta'. \tag{A.81}$$

Note that (A.73) and (A.80) are duals. Similarly, (A.75) and (A.78) can be obtained by using the duality property of electric and magnetic field interchange and the replacement of the surface impedance with its reciprocal.

A.5.1 Field Observables

Once the boundary integral equations have been solved, the scattered electric field is defined by the surface integral representation

$$\mathbf{E}_s(x,z) = -\frac{k_0\eta_0}{4}\int H_0^{(1)}(k_0\Delta\rho)\mathbf{a}(\zeta')d\zeta'$$
$$-\frac{\eta_0}{4k_0}\int \nabla\nabla H_0^{(1)}(k_0\Delta\rho)\cdot\mathbf{a}(\zeta')d\zeta'$$
$$+\frac{i}{4}\int \nabla H_0^{(1)}(k_0\Delta\rho)\times\mathbf{b}(\zeta')d\zeta'. \tag{A.82}$$

By substituting the explicit forms of the surface currents the scattered field for horizontal polarization reduces to

$$E_y^s(x,z) = -\int \frac{i}{4}H_0^{(1)}(k_0\Delta\rho)\frac{\partial E_y}{\partial n_0}\left(\zeta'\right)d\zeta'$$
$$+\int \frac{i}{4}\frac{\partial H_0^{(1)}}{\partial n_0'}(k_0\Delta\rho)E_y\left(\zeta'\right)d\zeta'. \tag{A.83}$$

For vertical polarization the $\nabla\nabla H_0^{(1)}$ term enters explicitly, whereby it is easier to use the equivalent dual relation

$$H_y^s(x,z) = -\int \frac{i}{4}H_0^{(1)}(k_0\Delta\rho)\frac{\partial H_y}{\partial n_0}\left(\zeta'\right)d\zeta'$$
$$+\int \frac{i}{4}\frac{\partial H_0^{(1)}}{\partial n_0'}(k_0\Delta\rho)H_y\left(\zeta'\right)d\zeta'. \tag{A.84}$$

Because the integrals must be evaluated for every value of x and z, however, further simplification is desirable. The standard approach uses the large argument form of the Green function, namely

$$\frac{i}{4}H_0^{(1)}(k\Delta\rho) \sim \frac{\exp\{ik\Delta\rho\}}{\sqrt{-i2\pi k\Delta\rho}}. \tag{A.85}$$

For completeness, the far-field result described in Section A.1 takes the form

$$\lim_{\rho\to\infty} F(x,z) = \widehat{F}(\mathbf{k}_\rho)\frac{\exp\{ik\rho\}}{\sqrt{-2\pi i k\rho}}, \tag{A.86}$$

where the vector $\mathbf{k}_\rho$ is the wave vector pointing in the direction of the observation point, and

$$\begin{aligned}
\widehat{F}(\mathbf{k}_\rho) &= \frac{1}{2}\int \frac{\partial F_y}{\partial n_0}(\zeta')\,\exp\{ik_\rho\cdot\Delta\boldsymbol{\rho}_s\}d\zeta' \\
&\quad -\frac{1}{2}\int (\mathbf{n}_0'\cdot\mathbf{k}_\rho)\,F_y(\zeta')\exp\{ik_\rho\cdot\Delta\boldsymbol{\rho}_s\}\,d\zeta', \tag{A.87}
\end{aligned}$$

with F and F_y replaced by the appropriate electric/magnetic source function and field component.

A.5.2   Highly Conducting Media

The boundary integral equations that apply within the medium that contains no impressed sources are effectively an implicit relation between the transverse field and its normal component. For a highly conducting medium,

$$k_1 = k_0\epsilon_1 = \sqrt{1 - j\sigma/\omega}, \tag{A.88}$$

with $\sigma/\omega \gg 1$. Thus, any term involving $\exp\{ik_1\gamma\}$ will decay exponentially unless γ is close to zero. Using (A.74) and (A.76) to establish explicit functional relationships between the transverse fields and their normal derivatives, it follows that

$$\frac{\partial F_y}{\partial N} + ik_0\beta_{H,V}F_y = 0, \tag{A.89}$$

where F_y represents the transverse field component for the specific polarization,

$$\beta_H = \eta_1\left[\sqrt{1+f'^2} + \frac{if''}{2k\sqrt{1+f'^2}}\right], \tag{A.90}$$

and $\beta_V = \beta_H^{-1}$. Note that for a perfectly conducting surface $\eta_1 \to \infty$, whereby the perfectly conducting boundary conditions, $E_y = 0$ for horizontal polarization and $\partial H_y/\partial N = 0$ for vertical polarization, are recovered. For a flat surface, β is the surface impedance defined as the ratio of transverse electric and magnetic field components.

The fact that the transverse electric field is small at the surface of a highly conducting boundary favors the MFIE both for horizontal polarization (A.75) and for vertical polarization (A.80). Upon substituting from the appropriate impedance boundary condition with $S(x) = \sqrt{1+f'^2(x)}$, the following BIEs, which can be solved directly, are obtained for horizontal polarization,

$$
\begin{aligned}
\frac{1}{2}\frac{\partial E_y}{\partial n_0}(\zeta) \;=\;& \frac{\partial E_{iy}}{\partial n_0}(\zeta) - \int \frac{i}{4}\frac{\partial H_0^{(1)}}{\partial n_0}(k_0\Delta\rho_s)\frac{\partial E_y}{\partial n_0'}\,d\zeta' \\
& -\frac{i}{k_0\eta_1}\int \frac{ik_0^2}{4}H_0^{(1)}(k_0\Delta\rho_s)\left(1+\frac{\partial f}{\partial x}\frac{\partial f}{\partial x'}\right)S^{-1}(x')\frac{\partial E_y}{\partial n_0'}\,d\zeta' \\
& -\frac{i}{k_0\eta_1}\int \frac{i}{4}\Omega(k_0,\zeta,\zeta')S^{-1}(x')\frac{\partial E_y}{\partial n_0'}\,d\zeta',
\end{aligned}
\tag{A.91}
$$

and for vertical polarization,

$$
\begin{aligned}
\frac{1}{2}H_y(\zeta) \;=\;& H_{iy}(\zeta) + \int \frac{i}{4}\frac{\partial H_0^{(1)}}{\partial n_0'}(k_0\Delta\rho_s)H_y(\zeta')\,d\zeta' \\
& -\frac{ik_0}{\eta_1}\int \frac{i}{4}H_0^{(1)}(k_0\Delta\rho_s)S^{-1}(x')H_y(\zeta')\,d\zeta'.
\end{aligned}
\tag{A.92}
$$

A.5.3   Numerical Solution

The following numerical solution for the horizontal polarization above a perfectly conducting boundary is taken from Rino and Kruger [26]. For horizontal polarization

$$
\begin{aligned}
\frac{\partial S}{\partial N}(x_j) \;=\;& 2\frac{\partial S^i}{\partial N}(x_j) - \frac{i}{2}\sum_{l=1}^{j-1}M_{jl}\frac{\partial S}{\partial N}(x_l) \\
& -2M_{jj}\frac{\partial S}{\partial N}(x_j),
\end{aligned}
\tag{A.93}
$$

where

$$
M_{jl} = \begin{cases} \dfrac{\frac{ik^2}{2}H_0^{(1)}(k\Delta r_{jl})\xi_{jl}}{k\Delta r_{jl}} & \text{for } j\neq l \\[2ex] \dfrac{\partial^2 f_j/\partial x^2}{1+(\partial f_j/\partial x)^2}\dfrac{\Delta x}{4\pi} & \text{for } j=l \end{cases},
\tag{A.94}
$$

$$
\xi_{jl} = \left[(x_j - x_l)^2 + (x_j - x_l)\,\partial f_j/\partial x\right]\Delta x,
\tag{A.95}
$$

and

$$
\Delta r_{jl} = \sqrt{(x_j - x_l)^2 + (f_j - f_l)^2}.
\tag{A.96}
$$

The source function is updated by the recursion

$$
\begin{aligned}
S(x_j) \;=\;& 2S(x_j) + \frac{i}{2}\sum_{l=1}^{j-1}M_{lj}S(x_l) \\
& +2M_{jj}S(x_j).
\end{aligned}
\tag{A.97}
$$

REFERENCES

1. Kung Chie Yeh and Chao-Han Liu. Radio wave scintillations in the ionosphere. *Proc. IEEE*, 70(4):324–360, 1982.

2. Jules Aarons. Global morphology of ionospheric scintillations. *Proc. IEEE*, 70(4):360–378, 1971.

3. Akira Ishimaru. Theory and application of wave propagation and scattering in random media. *Proc. IEEE*, 65(7):1030–1061, 1977.

4. Akira Ishimaru. *Wave Propagation and Scattering in Random Media, Volume I: Single Scattering and Transport Theory*. Academic Press, New York, 1978.

5. Akira Ishimaru. *Wave Propagation and Scattering in Random Media, Volume II: Multiple Scattering, Turbulence, Rough Surfaces, and Remote Sensing*. Academic Press, New York, 1978.

6. E. J. Fremouw, R. L. Leadabrand, R. C. Livingston, M. D. Cousins, C. L. Rino, B. C. Fair, and R. A. Long. Early results from the DNA wideband satellite experiment–complex-signal scintillation. *Radio Science*, 13(1):167–187, 1978.

7. V. I. Tatarskii, A. Ishimaru, and V. U. Zavorotny. *Wave Propagation in Random Media (Scintillation)*. SPIE, Bellingham, Washington, 1992.

8. R. H. Harrington. *Time-Harmonic Electromagnetic Fields*. McGraw-Hill, New York, 1961.

9. K. C. Yeh and C. H. Liu. *Theory of Ionospheric Waves*. Academic Press, New York, 1972.

10. J. A. Kong. *Electromagnetic Wave Theory: Second Edition.* John Wiley & Sons, New York, 1986.

11. Robert E. Collin. *Field Theory of Guided Waves.* IEEE Press, Piscataway, New Jersey, 1991.

12. When Cho Chew. *Waves and Fields in Inhomogeneous Media.* New York, Van Nostrand Reinhold, 1990.

13. Hoc D. Ngo and Charles L. Rino. Wave scattering functions and their application to multiple scattering problems. *Waves in Random Media*, 3(3):199–210, 1993.

14. Hoc D. Ngo and Charles L. Rino. Applications of the mutual interaction method to a general class of two-scatterer systems, Part I: Two discrete scatterers. *Waves in Random Media*, 5:89–105, 1995.

15. Hoc D. Ngo and Charles L. Rino. Applications of the mutual interaction method to a general class of two-scatterer systems, Part II: Scatterer near an interface. *Waves in Random Media*, 5:107–123, 1995.

16. Charles L. Rino and Hoc D. Ngo. Low-frequency acoustic scatter from subsurface bubble clouds. *J. Acoust. Soc. Am.*, 90(1):406–415, 1991.

17. H. G. Booker, J. A. Ratcliffe, and D. H. Shinn. Diffraction from an irregular screen with applications to ionospheric problems. *Proc. Camb. Phil. Soc.*, 242 A:579–607, 1950.

18. H. G. Booker and W. E. Gordon. A theory of radio scattering in the troposphere. *Proc. IRE*, 38:401–412, 1950.

19. B. H. Briggs and I. A. Parkin. On the variation of radio star and satellite scintillations with zenith angle. *J. Atmos. and Terr. Phys.*, 25:339–365, 1962.

20. J. P. Mercier. Diffraction by a screen causing large random phase fluctuations. *Proc. Camb. Phil. Soc.*, 58:382–400, 1961.

21. S. A. Bowhill. Statistics of a radio wave diffracted by a random ionosphere. *J. Res. Nat. Bur. Stand. D*, 242 A:275–292, 1961.

22. K. G. Budden. The amplitude fluctuations of the radio wave scattered from a thick ionospheric layer with weak irregularities. *J. Atmos. and Terr. Phys.*, 27:155–172, 1964.

23. K. G. Budden. The theory of the correlation of amplitude fluctuations of radio signals at two frequencies simultaneously scattered by the ionosphere. *J. Atmos. and Terr. Phys.*, 27:883–897, 1965.

24. C. L. Rino and E. J. Fremouw. The angle dependence of singly scattered wavefields. *J. of Atmos. Terr. Phys.*, 39:859–868, 1977.

25. Michael C. Kelley. *The Earth's Ionosphere Plasma Physics and Electrodynamics Second Edition.* Elsevier, Inc., London, 2008.

26. C. L. Rino and Valerie R. Kruger. A comparison of forward-boundary-integral and parabolic-wave-equation propagation models. *IEEE Proc. Antennas Propagat.*, 49(4):574–582, 2001.

27. David A. de Wolf. Backscatter corrections to the parabolic wave equation. *J. Opt. Soc. Am. A*, 6(2):174–179, 1989.

28. M. D. Collins. Benchmark calculations for higher-order parabolic equations. *J. Acoust. Soc. Am*, 87(4):1535–1538, 1990.

29. M. Born and E. Wolf. *Principles of Optics*. Cambridge University Press, 1999.

30. Mireille Levy. *Parabolic Equation Methods for Electromagnetic Wave Propagation*. Institute of Electrical Engineers, United Kingdom, 2000.

31. Albert D. Wheelon. *Electromagnetic Scintillation 1. Geometrical Opatics*. Cambridge University Press, 2001.

32. V. I. Tatarskii. *The Effects of the Turbulent Atmosphere on Wave Propagation*. Nat. Tech. Inform. Service, Springfield, VA, 1971.

33. Yu N. Barabenenkov, Yu A. Kravtsov, S. M. Rytov, and V. I. Tararskii. Status of the theory of propagation of waves in a randomly inhomogeneous medium. *Sov. Phys. Uspenkhi*, 13:551–575, 1971.

34. L. C. Lee. Wave propagation in a random medium: A complete set of the moment equations with different wavenumbers. *J. Math. Phys.*, 15(9):1431–1436, 1974.

35. E. K. Smith and S. Weintraub. The constants in the equation for atmospheric refractive index at radio frequencies. *Proc. IEEE*, 41:1025–1037, 1953.

36. B. R. Bean and G. D. Thayer. Models of the atmospheric radio refractive index. *Proc. IRE*, 48:740–754, 1959.

37. R. Paulus. Proceedings of the Electromagnetic Propagation Workshop. Technical document, Naval Command, Control, and Ocean Surveillance Center RDT&E Division, December 1995.

38. K. G. Budden. *The Propagation of Radio Waves*. Cambridge University Press, Cambridge, 1985.

39. Kenneth Davies. Recent progress in satellite radio beacon studies with particular emphasis on the ATS-6 radio beacon experiment. *Space Science Reviews*, 25(4):357–430, 1980.

40. L. J. Nickisch. A power-law spectral density model of total electron content structure in the polar region. *Radio Science*, 39:–, 2004.

41. Athanasios Papoulis. *Probability, Random Variables, and Stochastic Processes*. McGraw-Hill, New York, 1965.

42. I. P. Shkarofsky. Generalized Turbulence Space-Correlation and Wave-Number Spectrum-Function Pairs. *Can. J. Phys.*, 46:2683–2153, 1968.

43. D. G. Singleton. Saturation and focusing effects in radio-star and satellite scintillations. *J. Atmos. and Terr. Phys.*, 32:187–208, 1970.

44. M. J. Sewell. *Maximum and Minimum Principles, A Unified Approach with Applications*. Cambridge University Press, Cambridge, 1987.

45. C. L. Rino, V. H.Gonzalez, and A. R. Hessing. Coherence bandwidth loss in transionospheric radio propagation. *Radio Science*, 16(2):245–255, 1982.

46. K. C. Yeh, C. H. Liu, and M. Y. Youakim. A theoretical study of the ionospheric scintillation behavior caused by multiple scattering. *Radio Science*, 10(1):97–106, 1975.

47. Marilyn Marians. Computed scintillation spectra for strong turbulence. *Radio Science*, 10(1):115–119, 1975.

48. V. H. Rumsey. Scintillation due to a concentrated layer with a power-law turbulence spectrum. *Radio Science*, 10(1):107–114, 1975.

49. V. I. Shishov. Diffraction of waves by a strongly refracting random phase screen. *Radio Phys. and Quant. Elect. Izvestiya Translation*, 14(1):70–75, 1971.

50. K. S. Gochelashvily and V. I. Shishov. Laser beam scintillation beyond a turbulent layer. *Optica Acta*, 18(4):313–320, 1971.

51. K. S. Gochelashvily and V. I. Shishov. Multiple scattering of light in a turbulent medium. *Optica Acta*, 18(10):767–777, 1971.

52. V. I. Shishov. Dependence of the form of the scintillation spectrum on the form of the spectrum of refractive-index inhomogeneities. *Radio Phys. and Quant. Elect. Izvestiya Translation*, 17(11):1287–1292, 1974.

53. K. S. Gochelashvily and V. I. Shishov. Saturated fluctuations in the laser radiation intensity in a turbulent medium. *Sov. Phys. JETP*, 39(4):605–609, 1974.

54. E. E. Salpeter. Interplanetary Scintillations I. Theory. *Astrophys. J.*, 147:433–448, 1967.

55. Johanan L. Codona, Dennis B. Creamer, Stanley M. Flatte, R. G. Frelich, and Frank S. Henyey. Solution for the fourth moment of waves propagating in random media. *Radio Science*, 21(6):929–948, 1986.

56. Stanley M. Flatte. The Schrodinger Equation in Classical Physics. *Am. J. Phys.*, 54(12):1088–1092, 1986.

57. B. J. Uscinski. Analytical solution of the fourth-moment equation and interpretation as a set of phase screens. *J. Opt. Soc. Am. A*, 2(12):2077–2091, 1985.

58. Alan M. Whitman and Mark J. Beran. Two-scale solution for atmospheric scintillation. *J. Opt. Soc. Am. A*, 2(12):2133–2143, 1985.

59. J. Gozani. Numerical solution for the fourth-order coherence function of a plane wave propagating in a two-dimensional Kolmogorovian medium. *J. Opt. Soc. Am. A*, 2(12):2144–2151, 1985.

60. Moshe Tur. Numerical solutions for the fourth moment of a finite beam propagating in a random medium. *J. Opt. Soc. Am. A*, 2(12):2161–2170, 1985.

61. C. H. Liu and K. C. Yeh. Pulse spreading and wandering in random media. *Radio Science*, 14(5):925–931, 1979.

62. Johanan L. Codona, Dennis B. Cremer, Stanley M. Flatte, R. G. Frelich, and Frank S. Henyey. Moment-equation and path-integral techniques for wave propagation in random media. *J. Math. Phys.*, 21(5):171–177, 1985.

63. Johanan L. Codona, Dennis B. Cremer, Stanley M. Flatte, R. G. Frehlich, and Frank S. Henyey. Two-frequency intensity cross spectrum. *Radio Sci.*, 27(1):805–814, 1985.

64. R. Mazar, J. Gozani, and M. Tur. Two-scale solution for the intensity fluctuations of two-frequency wave propagation in a random medium. *J. Opt. Soc. Am. A*, 2(12):2152–2160, 1985.

65. J. M. Martin and Stanley M. Flatte. Intensity images and statistics from numerical simulation of wave propagation in 3-d random media. *Applied Optics*, 27(11):2111–2126, 1988.

66. J. M. Martin and Stanley M. Flatte. Simulation of point-source scintillation through three-dimensional random media. *J. Opt Soc. Am A*, 7(5):838–847, 1988.

67. Nikolay N. Zernov, V. E. Gherm, and Hal J. Strangeways. On the effects of scintillation of low-latitude bubbles on transionospheric paths of propagation. *Radio Science*, 44(10):1029–1038, 2009.

68. A. A. Bitjukov, V. E. Gherm, and Nikolay N. Zernov. On the solution of markov's parabolic equation for the second-order space frequency and position coherence function. *Radio Science*, 37(4):1066–, 2001.

69. Vadim E. Gherm, Nikolay N. Zernov, Sandro M. Radicella, and Hal J. Strangeways. Propagation model for signal fluctuations on transionsopheric radio links. *Radio Science*, 35(5):1221–1232, 2000.

70. C. L. Rino. A power law phase screen model for ionospheric scintillation 2. Strong scatter. *Radio Science*, 14(6):1147–1155, 1979.

71. Stanley M. Flatte and James S. Gerber. Irradiance-variance behavior by numerical simulation for plane-wave and spherical-wave optical propagation through strong turbulence. *J. Opt. Soc. Am.*, 17(6):1092–1097, 2000.

72. J. A. Ratcliffe. Some aspects of diffraction theory and their application to the ionosphere. *R. Phys. Soc. Progr. Phys.*, 19:188–267, 1954.

73. M. Nakagami. The m-distribution: A general formula of intensity distribution of rapid fading. In *Statistical Methods in Radio Propagation*. Pergamon, New York, 1960.

74. Roger Dashen. Distribution of intensity in a multiply scattering medium. *Optics Letters*, 10(4):110–112, 1984.

75. Roger Dashen, Guang-Yu Wang, Stanley M. Flatte, and Charles Bracher. Moments of intensity and log intensity: New asymptotic results for waves in power-law media. *J. Opt. Soc. Am. A*, 10(6):1233–1242, 1993.

76. E. Jakeman. Speckle statistics with a small number of scatterers. *Optical Eng.*, 23(4):453–461, 1984.

77. M. V. Berry. Focusing and twinkling: Critical exponents from catastrophes in non-Gaussian random short waves. *J. Phys. A: Math. Gen.*, 10(12):2061–2081, 1977.

78. C. L. Rino, R. T. Tsunoda, J. Petriceks, R. C. Livingston, M. C. Kelley, and K. D. Baker. Simultaneous rocket-borne beacon and in situ measurements of equatorial spread f–intermediate wavelength results. *J. Geophys. Res. A*, 86(A4):2411–2419, 1981.

79. M. C. Kelley, R. C. Livingston, C. L. Rino, and R. T. Tsunoda. The vertical wave number spectrum of topside equatorial spread F: Estimates of backscatter levels and implications for a unified theory. *J. Geophys. Res. A*, 87(A7):5217–5221, 1982.

80. S. J. Franke, C. H. Liu, and D. J. Fang. Multifrequency study of ionospheric scintillation at Ascension Island. *Radio Science*, 19(3):695–706, 1984.

81. Emanoel Costa and Santimay Basu. A radio wave scattering algorithm and irregularity model for scintillation predictions. *Radio Science*, 37(3):1046–, 2002.

82. C. L. Rino and J. Owen. The structure of localized nighttime auroral zone scintillation enhancements. *J. Geophys. Res. A*, 85(A6):2941–2948, 1981.

83. W. W. Berning. Charge densities in the ionosphere from radio Doppler data. *J. of Meteror.*, 8(3):175–183, 1951.

84. C. L. Rino and R. C. Livingston. On the analysis and interpretation of spaced-receiver measurements of transionospheric radio waves. *Radio Science*, 17(4):845–854, 1982.

85. B. H. Briggs and M. Spencer. Horizontal movements in the ionosphere. *Rep. Prog. Phys.*, 17:245–280, 1954.

86. J. W. Armstrong, W. A. Coles, M. Kojima, and B. J. Rickett. Observations of field-aligned density fluctuations in the solar wind. *Astrophys. J.*, 358:685–692, 1990.

87. C. L. Rino and J. Owen. The time structure of transionospheric radio wave scintillation. *Radio Science*, 15(3):479–489, 1980.

88. K. C. Yeh and C. C. Yang. Mean arrival time and mean pulse width of signals propagating through a dispersive and random medium. *IEEE Trans. Ant. and Prop.*, 25(5):711–713, 1977.

89. K. C. Yeh and K. C. Liu. An investigation of temporal moments of stochastic waves. *Radio Science*, 12(5):671–680, 1977.

90. K. C. Yeh and K. C. Liu. Ionospheric effects on radio communication and ranging pulses. *IEEE Trans. Ant. and Propagat.*, 27(6):747–751, 1979.

91. D. L. Knepp. Analytic solution for the two-frequency mutual coherence function for spherical wave propagation. *Radio Science*, 18(4):535–549, 1983.

92. D. L. Knepp and Leon A. Wittwer. Simulation of wide bandwidth signals that have propagated through random media. *Radio Science*, 19(1):303–318, 1984.

93. Gilbert Strang and Kai Borre. *Linear Algebra, Geodesy, and GPS*. Wellesley-Cambridge Press, Mass., 1997.

94. David A. Vallado, Paul Crawford, Richard Hujsak, and T. S. Kelso. Revisiting spacetrack report No. 3: Rev 1. Technical Report AIAA 2006-6753-Rev1, American Institute of Aeronautics and Astronautics, 2006.

95. H. D. Craft and L. H. Westerlund. Scintillation at 4 and 6 GHz caused by the ionosphere. *Amer. Inst. of Aero. and Astr.*, pages 172–179, 1972.

96. Dennis C. Ghiglia and Mark D. Pritt. *Two-Dimensional Phase Unwrapping Theory, Algorithms, and Software*. John Wiley and Sons, Inc., New York, 1998.

97. P. Flandrin. Wavelet analysis and synthesis of fractional Brownian motion. *IEEE Trans. Info. Th.*, 38:910–917, 1992.

98. P. Abry, P. Goncalves, and P. Flandrin. Wavelets spectrum analysis and 1/f processes. In *Wavelets and Statistics*. Springer Verlag, 1995.

99. L. Hudgins, C. A. Friehe, and M. E. Mayer. Wavelet transforms and atmospheric turbulence. *Phys. Rev. Letters*, 71:3279–3282, 1993.

100. Robert S. Kennedy. *Fading Dispersive Communication Channels*. John Wiley and Sons, Inc., New York, 1969.

101. D. Slepian and H. O. Pollak. Prolate spheroidal wave functions, Fourier analysis and uncertainty-I. *Bell Syst. Tech. J.*, 40:43–63, 1961.

102. D. L. Knepp. Multiple phase-screen calculation of the temporal behavior of stochastic waves. *Proc. IEEE*, 71(6):722–737, 1983.

103. Mark A. Richards. *Fundamentals of Radar Signal Processing*. Mc Graw-Hill, New York, 2005.

104. R. L. Bogusch, F. W. Guigliano, D. L. Knepp, and A. H. Michelet. Frequency selective propagation effects on spread-spectrum receiver tracking. *Proc. IEEE*, 69:787–796, 1981.

105. C. L. Rino. Numerical computations for a one-dimensional power law phase screen. *Radio Science*, 15(1):41–47, 1980.

106. D. L. Knepp. Aperture antenna effects after propagation through strongly disturbed random media. *IEEE Proc. Antennas Propagat.*, 33(10):1074–1083, 1985.

107. A. Bhattacharyya, K. C. Yeh, and S. J. Franke. Deducing turbulence parameters from transionospheric scintillation measurements. *Space Sci. Rev.*, 61:335–386, 1992.

108. Albert D. Wheelon. *Electromagnetic Scintillation 2. Weak Scattering*. Cambridge University Press, 2001.

109. Robert K. Crane. Ionospheric scintillation. *Proc. IEEE*, 68(2):180–199, 1977.

110. Jules Aarons. 50 years of radio-scintillation observations. *IEEE Antennas Propagat. Mag.*, 39(6):7–12, 1997.

111. Thomas P. Yunck, Chao-Han Liu, and Randolph Ware. A history of GPS sounding. *Terr., Atmos., and Oc. Sci.*, 11(1):1–25, 2000.

112. Paul A. Bernhardt and Carl L. Siefring. New satellite-based systems for ionospheric tomography and scintillation region imaging. *Radio Science*, 41:1–14, 2006.

113. Paul A. Bernhardt, Carl L. Siefring, Ivan J. Galysh, Thomas F. Rodilosso, Douglas E. Koch, Thomas L. MacDonald, Matthew R. Wilkens, and G. Paul Landis. Ionospheric applications of the scintillation and tomography receiver in space (CITRIS) mission when used with the DORIS radio beacon network. *J. Geod.*, 80:473–692, 2006.

114. Paul A. Bernhardt, Carl L. Siefring, Patrick A. Roddy, and Donald E. Hunton. Comparisons of equatorial irregularities measurements from C/NOFS: TEC using CERTO and CITRIS with in-situ Plasma Density. *Geophys. Res. Let.*, 36:1–5, 2009.

115. Victor Twersky. Signals, scatterers, and statistics. *IEEE Proc. Antennas Propagat.*, 11(6):363–364, 1963.

116. G. Daniel Dockery and James R. Kuttler. An improved impedance boundary algorithm for Fourier split-step solutions of the parabolic wave equation. *IEEE Trans. Antennas Propagat.*, 44(12):1592–1599, 1996.

117. Charles L. Rino. Double-passage radar cross section enhancements. *Radio Science*, 29(2):495–501, 1994.

118. D. L. Knepp and H.L. Houpis. Altair VHF/UHF observations of multipath and backscatter enhancement. *IEEE Proc. Antennas Propagat.*, 39(4):528–534, 1991.

119. J. R. Guerci. *Space-Time Adaptive Processing for Radar.* Artech House, Inc., Norwood, MA, 2003.

120. Nagayoshi Morita, Nobuaki Kumagai, and Joseph R. Mautz. *Integral Equation Methods for Electromagnetics.* Artech House, Boston, 1990.

121. T. B. A. Senior and J. L. Volakis. *Approximate Boundary Conditions in Electromagnetics.* Institute of Electrical Engineers, United Kingdom, 1995.

122. G. R. Valenzuela. Scattering of electromagnetic waves from a tilted slightly rough surface. *Radio Science*, pages 1057–1066, 1986.

123. Leung Tsang, Jin Au Kong, and Robert T. Shin. *Theory of Microwave Remote Sensing.* John Wiley & Sons, New York, 1985.

124. Robert J. Adams and Gary S. Brown. A rapidly convergent iterative method for two-dimensional closed body scattering problems. *Microwave and Optical Technology Letters.*, 20(3):179–183, 1999.

125. Charles L. Rino, Kenneth Doniger, and Raul Martinez. The method of ordered multiple interactions for closed bodies. *IEEE Trans. Antennas Propagat.*, 51(9):2327–2334, 2003.

126. Charles L. Rino and Hoc D. Ngo. Forward propagation in a half-space with and irregular boundary. *IEEE Trans. Antennas Propagat.*, 45(9):1340–1347, 1997.

127. Editor-Gary S. Brown. Special issue on low-grazing-angle backscatter from rough surfaces. *IEEE Proc. Antennas Propagat.*, 46(1):1–161, 1998.

INDEX

Printed and bound by CPI Group (UK) Ltd, Croydon, CR0 4YY

16/04/2025

14658351-0001